Atlas of Wildlife

Illustrated by
Adrian Williams & **David Nockels**
Consultant: Dr. Maurice Burton

Maps by Geographical Projects London

Jacqueline Nayman
Atlas of Wildlife

The John Day Company   New York

Geographical Director: **Shirley Carpenter**
Editor: **Geoffrey Rogers**
Art Director: **Frank Fry**
Design: **Nigel Talbot**

The John Day Company, 257 Park Avenue South, New York, N.Y. 10010

An **Intext** Publisher

Published in Canada by Longman Canada Limited

Library of Congress Cataloging in Publication Data
Nayman, Jacqueline
    Atlas of Wildlife
    1. Zoogeography   I. Title.
QL101.N38   591.9   78-38034

Printed and bound in Belgium by Brepols, Turnhout

Contents

Introduction   11

Chapter: 1   Animal Distribution   12

2   The Palaearctic Region   22

3   The Nearctic Region   40

4   The Neotropical Region   52

5   The Ethiopian Region   68

6   The Oriental Region   84

7   The Australian Region   94

8   Antarctica   104

9   Islands   108

Wildlife Parks of the World   118

Index   121

Introduction

ATLAS OF WILDLIFE is about wild animals and where they live. Although we may know the size and general appearance of many of the most familiar wild animals, when we are asked to say exactly where such animals live we often answer vaguely, "Oh, somewhere in Africa or South America." ATLAS OF WILDLIFE dispels such vagueness in an immediate and compelling way. By turning to the large, full-colour relief maps, on which animals are plotted with easily recognizable symbols, the reader is able to trace at a glance the distribution of wild animals in all regions of the world.

ATLAS OF WILDLIFE is organized on a simple and effective plan that promotes the reader's understanding of animal distribution. Chapter 1 explains how animals are thought to have reached their present-day homes and introduces the six zoogeographical regions, or realms—the term naturalists use to describe areas of the world that have distinctive animal populations. These regions are then discussed in turn in Chapters 2 to 7. The text describes in a clear, nontechnical way the characteristic animals of each region, relating their distribution to the climatic and vegetational zones within that region, and, where applicable, to the disruptive effect of man. As well as fully captioned illustrations and detailed distribution maps, these six chapters also have summary charts that list the main groups of animals found in each region and indicate whether or not they occur in other regions. Chapter 8 describes the animals that live in and around Antarctica, and Chapter 9 features the unique animals of four widely separated island groups.

The distribution maps of ATLAS OF WILDLIFE feature mainly mammals (wolves, deer, elephants, bears, seals, etc.), but a few amphibians, reptiles, and birds are also included. The symbols have been plotted according to the known distribution of each animal, taking into account the latest information available on those animals whose range is decreasing rapidly. Where symbols are spaced out evenly this usually means that the animal is found in a continuous distribution over that area, whereas a single, isolated symbol usually represents a restricted distribution. In some cases, however, spaced-out symbols represent small populations at distinct points over a wide area. One example of this is the Mediterranean monk seal, which is known to occur only at selected spots around the Mediterranean coast. Where a special interpretation of this sort is required, a caption appears on the map or on the preceding page. Unless otherwise stated, each symbol represents all types of the animal in question. The sloth symbol, for example, refers to both the two-toed and the three-toed variety, and symbols appear wherever one or both of these varieties are found.

Although many wild animals are near extinction, ATLAS OF WILDLIFE closes on an optimistic note by presenting a map of the world's major wildlife parks.

1 Animal Distribution

Indo-China

W hy is it that some animals occur only in certain areas of the world, kangaroos, for example, in Australia and New Guinea, or sloths in South America, whereas others are widespread? As a first step in answering this question, many naturalists of the 1800's proposed ways of dividing the world into regions according to the animals that lived in them. One of these naturalists was the Englishman Alfred Russel Wallace. In 1876, he divided the world into six regions, sometimes called realms. These zoological, or zoogeographical, regions were modelled very closely on those proposed by P. L. Sclater, the Secretary of the Zoological Society of London, who, in 1857, divided up the world according to the distribution of birds. Wallace, however, had in mind a set of regions that would apply to all animals. It is a tribute to the thoroughness of Sclater and Wallace that their regions are still generally accepted by today's naturalists. The modern map of the zoogeographical regions shown on pages 16–17 differs only slightly from Wallace's original version.

The division of the world into zoogeographical regions was a useful device in the study of animal distribution, a device that Wallace found necessary in the light of an event that had taken place some years earlier, in 1858, in London. This was the joint presentation by Charles Darwin and Wallace of the theory of evolution by natural selection—a theory each had worked on and developed independently. Before this theory was made known, naturalists thought that animals had been created in the place most suitable for them. This idea, though seemingly logical and straightforward, had its problems. It was, for example, difficult to explain why animals introduced into a new country thrived and spread so well. As the area so obviously suited the newcomer, why had it not been created there?

Wallace's line between the Oriental and Australian zoogeographical regions. Wallace started to classify the islands of the Malay archipelago as either Oriental or Australian in 1856, after travelling from Bali to Lombok and noticing the sharp difference in the animal populations of these two islands. (See page 14 for the names of the animals shown.)

Philippine Islands

Wallace's Line

Mindanao

Borneo

Bali

Lombok

Australia

According to the theory of evolution all present-day animals have evolved over millions of years from other animals. All animals had common ancestors, although naturalists do not always know what they looked like. The theory of evolution cut across all the ideas held before 1858 and brought with it the need for a different explanation for the distribution of animals.

One of the basic facts known by Wallace and his fellow naturalists was that most animals reproduce themselves at such a fast rate that they tend to spread. This spread is limited by several factors, including climate, vegetation, other animals, and physical barriers. The cold climate prevents the spread of reptiles into the far South and far North, and the relatively warm conditions in temperate and tropical areas prevent the polar bear from spreading south from its Arctic home.

The limiting effect of vegetation is often closely linked with the effects of climate, but there are certain plant-eating animals that have a very restricted distribution because they can feed on only one type of plant. Good examples of this are the koala, which feeds on the leaves of the eucalyptus trees found in eastern Australia, and the giant panda of the Chinese bamboo forests.

Interaction between animals is a complicated process. At its simplest one animal may prevent the spread of another because they are both competing for the same, limited food supply.

The most obvious factor limiting the spread of animals is a physical barrier. Mountains, deserts, and wide rivers can all stop the spread of certain animals, but the sea is the most effective barrier to all but marine animals. However, some animals can fly over the seas—bats, for example, which are consequently found the world over. Some land animals, such as tapirs, are strong swimmers, and so, too, surprisingly, are pigs. Tortoises are thought to be able to float for great distances, and many small animals may "raft" from land to land on driftwood or on islands of debris swept out to sea by flooding rivers.

Although land animals may have reached certain islands by swimming or floating across the sea, they could not have covered the vast distances from one continent to another in this way. Naturalists are therefore forced to the conclusion that at some time in the earth's history there must have been land connections between some of the continents.

Scientists know that at various times in the earth's history the level of the seas has been considerably lower than it is now. During the Ice Age that began about one million years ago, for example, a large amount of sea water was locked up in vast ice sheets. At such times shallow seas would have dried up and certain continents would have been joined by land. The Nearctic and Palaearctic regions were probably joined at the Bering Strait, the narrow stretch of sea that now separates Alaska and Siberia.

Land bridges such as this may explain the distribution of certain animals (for example, bison probably spread from the Palaearctic to the Nearctic region at this time) but they cannot account for similarities in the animals of more distant continents, which could not have been joined during the Ice Age. How is it that fresh-water lungfish and side-necked turtles are found in

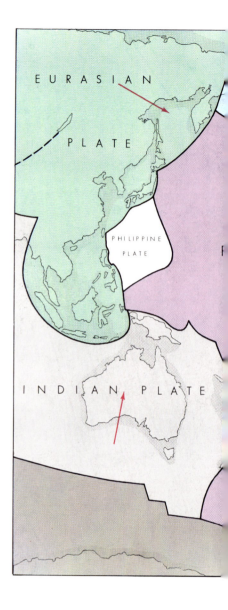

The animals illustrated on the opening page occur in the Oriental and Australian regions, not necessarily in the area shown on the map. The Oriental region animals shown are (from top to bottom): Lady Amherst's pheasant, a tarsier, a Philippine tree-shrew, and an Asiatic chevrotain. The Australian region animals are (from top to bottom): a white-throated tree kangaroo, a sulphur-crested cockatoo, and a common native cat.

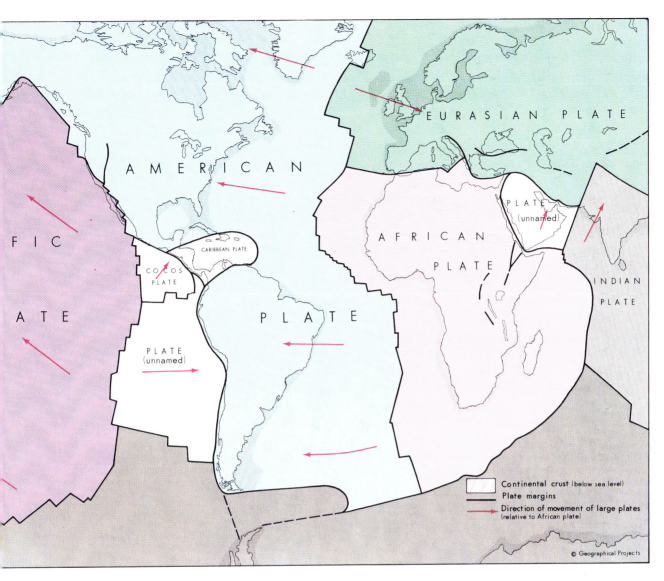

Scientists today consider that the surface of the Earth is divided into rigid plates that are moving over a molten or semi-molten layer beneath them. Six major plates and several smaller ones are shown on the world map above. By measuring the speed at which the plates move and by calculating the age of rocks in various places in the world, scientists have been able to work out the positions of the continents as they were many millions of years ago. The results of their studies have helped naturalists to explain certain mysteries in the present-day distribution of animals.

South America, Africa, and Australia? Equally puzzling is the fact that the only living relation of the New Zealand frog, *Leiopelma*, with its tail-wagging muscles but no tail, is found in the streams of North America. And how has one type of lizard come to live in both Madagascar and the Neotropical region? These were some of the mysteries that faced naturalists until they began to accept a theory that was first proposed in the early 1900's.

In 1915 a German scientist, Alfred Wegener, published a book, *The Origin of Continents and Oceans*, in which he put forward the theory that the continents were at one time joined together and had slowly drifted apart to their present positions. This theory, known as the theory of Continental Drift, was not generally accepted by scientists until the early 1960's, when the study of the sea floor together with other research work revealed evidence in support of the idea. Marine scientists found that molten rock was pouring out along the centre of the underwater mountain ranges, or ridges, that had been discovered in many of the world's oceans.

One of these ridges, the Mid-Atlantic Ridge, runs north to

THE ZOOGEOGRAPHICAL REGIONS OF THE WORLD

Projection: Gall

Scale: 1:46,100,000 equatorial scale
Miles

Kilometres

The Palaearctic & Nearctic regions are sometimes combined as the Holarctic region.

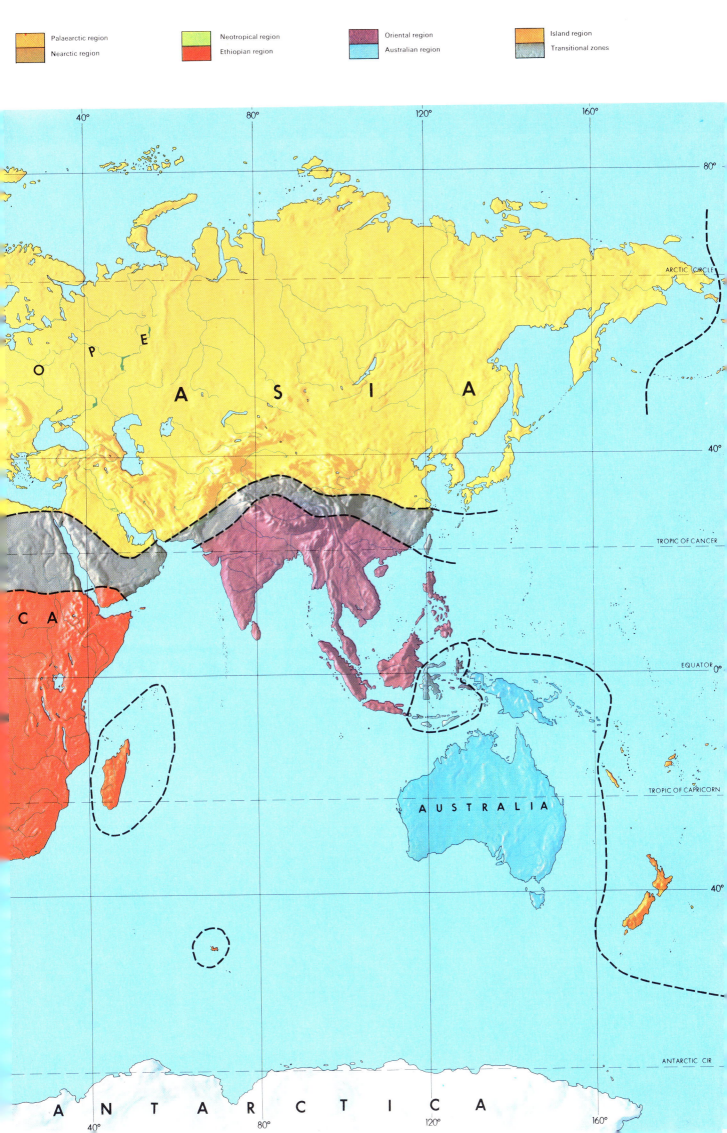

Palaearctic region
Nearctic region

Neotropical region
Ethiopian region

Oriental region
Australian region

Island region
Transitional zones

O P E

A S I A

C A

AUSTRALIA

ANTARCTICA

ARCTIC CIRCLE

40°

TROPIC OF CANCER

EQUATOR 0°

TROPIC OF CAPRICORN

40°

ANTARCTIC CIR

40°   80°   120°   160°

80°

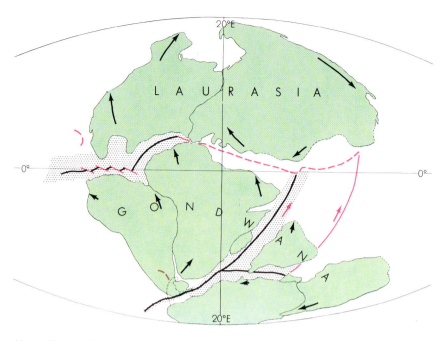

Above: The world as it may have looked 180 million years ago, about 20 million years after the single land mass of Pangaea first began to break up. The northern group of continents, Laurasia, has separated from the southern group, Gondwana, which is itself splitting apart. The stippled areas represent new ocean floor created as crustal material spreads from rifts. Some of the ancestors of modern fish, amphibians, and reptiles had evolved by this time and each fragment of the disintegrating Gondwana carried a share of these animals. This explains why such animals as fresh-water lungfish and side-necked turtles are shared by the distant continents of South America, Africa, and Australia.

Below: The world 65 million years ago. The southern continents have continued to drift apart, with South America, Africa, India, and Madagascar breaking free from the combined land mass of Antarctica and Australia. Laurasia is still a super-continent, but the part destined to become North America is moving westward and the shape of Greenland is becoming clear. Mammals were evolving at this time and since North America and Eurasia were joined, these mammals were able to spread throughout this land mass. This is one of the reasons why the mammals of the Nearctic and Palaearctic regions are very similar. As the top map shows, Gondwana began to split up many millions of years earlier and the mammals living on this super-continent became isolated from each other as the southern continents drifted apart. Once isolated, these mammals evolved in completely different ways. This explains why the mammals of the Neotropical, Ethiopian, and Australian regions are today so different from each other.

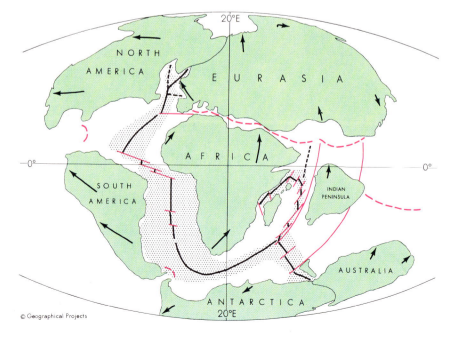

—————— Rifts

– – – – Tethyan Trench

⇒⇒⇒⇒ Zones of slipping along plate boundaries

——▶ Direction of movement of continents since drift began

© Geographical Projects

south midway between the American continents and the coastline of Europe and Africa. By using sensitive measurement techniques scientists have discovered that the final result of the slow eruption of rock along the Mid-Atlantic Ridge is that the continents are being pushed apart at a rate of one centimetre a year.

Many of today's scientists believe that the moving sea floor is made up of six large and several smaller rigid plates. They are pushed along, the scientists suggest, as new crust material surges out of the Earth at their trailing edges (i.e. at the mid-oceanic ridges). Unless the Earth is expanding, this extra material must be absorbed in some way and the scientists consider that this happens where the leading edges of the plates meet.

Where two plates of ocean crust meet, one over-rides the other and the lower plate descends to be destroyed in the molten interior of the Earth. Where this happens, a deep ocean trench appears and volcanoes rise, often forming volcanic islands. Some scientists think that at one time the Earth may have been covered by a single ocean, there being no dry land anywhere. As the plates of ocean crust moved and over-rode each other, they suggest, chains of islands were formed by volcanic action and these may have been starting points of our present continents.

However, not all plates are entirely covered by sea, some also carry continental crust. Where an ocean plate meets a plate with a continent on its leading edge, as seems to be happening on the west coast of South America, the ocean plate dips towards the centre of the earth. A trench forms close to the coast and earthquakes and volcanic action occur. In this case, however, the volcanic action happens on the land and as a result mountains are formed. The sea floor sediment on the plate that is dipping into the earth is scraped off on to the continent and helps in the formation of these mountains. The Andes mountains of South America were probably formed in this way. In fact many of the world's mountain ranges are situated near coast-lines and these too may have been formed in a similar way. Where mountain ranges occur inland (the Urals, for example) it seems probable that they mark the boundary of an ancient sea that no longer exists.

If one continental plate collides with another, neither can dip down and vast mountain ranges are thrown up. Scientists think that the Himalayas were formed in this way when the plate carrying India collided with Asia.

By studying the rate at which new crust material is produced in mid-oceanic ridges and by dating the rocks of the sea floor, scientists have been able to work out the approximate positions of the continents at various times in the Earth's history. Thus, scientists calculate that about 180 million years ago North America, Greenland, Europe, and Asia were one super-continent, which they have named Laurasia [Laur(entian) + (Eur)asia]. At about the same time Antarctica, Australia, New Zealand, India, Africa, and South America were joined to form a larger super-continent that scientists call Gondwana (from a district in India inhabited by a people called Gonds). These two super-continents were separated by the ancient Tethys ocean. It is possible that over 200 million years ago Laurasia and Gond-wana were joined as a single land mass, Pangaea ("all lands"), and that even earlier there were different groupings of continents.

Over a long period of time Pangaea divided into Laurasia and Gondwana and these super-continents moved away from each other and, in their turn, broke up. Africa and India moved northwards, eventually obliterating the Tethys ocean, and became joined to Asia. North and South America moved west-wards, the Atlantic Ocean forming as they did so, and then ultimately joined together. Finally Australia and New Zealand separated from Antarctica and moved northwards.

The theory that the continents of the world were once joined together certainly helps to explain certain puzzling facts in the present-day distribution of animals. Why, for example, are the animals of the two northern regions (Palaearctic and Ne-arctic) so similar, when the mammals, particularly, of three of the southern regions (Neotropical, Ethiopian, and Australian) are so dissimilar? It seems probable that animals have been able to spread between the two northern regions until comparatively recent times, due to the land connections at the Bering Strait already mentioned. The similarity of the animals of these two regions may also be explained by the fact that the portions of Laurasia seem to have split up and moved away from each other at a later date than parts of Gondwana, which gave rise to the southern continents.

Why are the mammals of the southern regions so dissimilar? Mammals evolved during and after the break-up of Laurasia and Gondwana. Once animals are isolated they evolve in different ways and since Gondwana began to break up first one would expect the mammals of the southern regions to show the greatest differences. The Neotropical region, for example, has sloths, armadillos, and anteaters—mammals that are found nowhere else in the world. The Ethiopian region alone has giraffes and hippopotamuses, and Australia has many unique pouched mammals, such as kangaroos and koalas.

What then of the similarity in some of the fresh-water fishes, amphibians, and reptiles of these southern continents? This, too, can be explained according to the theory of Continental Drift. The fresh-water fishes, amphibians, and reptiles evolved before the disintegration of Gondwana and would have presumably spread throughout this land mass. Thus, when South America, Africa, Australia, and New Zealand broke free, they would all have taken with them a share of these animals. This theory is supported by the recent discovery of fossil reptiles in Antarctica (originally part of Gondwana) similar to others found in Africa, India, and China.

The Ice Age that began about one million years ago also had a profound effect on the distribution of animals, especially those in the northern regions. Most of the Northern Hemi-sphere became covered in a great sheet of ice. This had the effect of lowering the temperature throughout the world and of reducing the level of the oceans, and most importantly it had a

Above: The maximum extent of glaciation in the Northern Hemisphere during the Pleistocene Ice Age. This Ice Age, which began over one million years ago and is considered by some scientists to have ended about 10,000 years ago, had a great effect on the development and distribution of animals, especially those of the Palaearctic and Nearctic regions. The ice advanced and retreated in four main cycles, causing world-wide changes in climate and alterations of sea level. A land bridge at the Bering Strait formed several times, allowing animals to migrate between Eurasia and North America. Bison, musk oxen, and moose probably spread into North America by way of this bridge. Many of the animals, such as sabre-toothed tigers and mammoths, that flourished in both regions during the Ice Age became extinct, either as a result of a change in physical conditions or at the hand of primitive man. Many other animals survived only in southern regions, such as Africa and southern Asia, which today have animal populations resembling those of Pleistocene times.

direct effect on the animals of the Nearctic and Palaearctic regions. Few animals could live on the ice-covered continents and so most were pushed south or became extinct. Mammoths and sabre-toothed tigers that had roamed the north for thousands of years became the victims of the Ice Age. Animals such as elephants, monkeys, and tapirs were forced to retreat and are now found only in the southern regions.

Man, too, has influenced the distribution of animals. He has introduced them into new continents, sometimes with disastrous effects. He has hunted them for food, clothing, and even for amusement. And he has destroyed their living places in order to grow crops and build towns. Of all the forces that have affected the distribution of animals, man's influence may yet prove to be the most decisive.

2 The Palaearctic Region

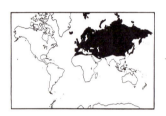

Europe, Africa north of the Sahara Desert, Asia Minor, the Middle East, Iran, Afghanistan, USSR, Mongolia, northern China, Korea, and Japan.

The Palaearctic is the largest of the zoogeographical regions. Within its wide-ranging boundaries the climate, and therefore the vegetation, varies enormously. A human being can freeze to death in Lapland or Siberia, or can die of heat-stroke in the Sahara or in the Arabian deserts.

The animals, too, are diverse, although few are unique to the region. The chart on page 38 shows that the Palaearctic region shares many of its animals with both the Ethiopian and Oriental regions. As these regions are connected by land this is not, perhaps, surprising. The region also has animals in common with North America, to which it was joined at several periods in the past.

Thus, most of the animals in Europe and Asia are found throughout the greater part of the world. Yet within the region itself, many of the animals have retreated under the advance of man and are now restricted in their distribution. Man has driven the large animals from their territories either by hunting them or by taking over the land for cultivation or building. Those that have survived have taken refuge in inaccessible areas, such as mountains or forests.

Man's disruptive influence on the wildlife of the Palaearctic region has taken effect only during the last thousand years. If we could have made a trip through Europe and Asia in, say, A.D. 1000, we would have seen many animals that are today struggling to survive, or are already extinct. On the plains of western Asia and eastern Russia we would have seen the now comparatively rare small, brown saiga antelope in vast herds that covered the steppes as far as the eye could see. Herds of wild horses, both the now extinct grey tarpan of western Asia and the now extremely rare Przewalski's horse of Mongolia,

A mature bull reindeer and young. Insulated from the cold by a thick, double-layered coat, reindeer dig through the winter snow with their broad, deeply cleft hooves to expose the lichens and moss on which they feed. Reindeer and the closely related caribou of the Nearctic region are the only deer in which both sexes grow antlers.

would have formed part of the landscape as we journeyed along. Our guide on the trip would have pointed out herds of wild asses and European bison. And he would almost certainly have shown us the largest mammal of the region, the European auroch, a giant wild ox, extinct since the 1500's. Preying on these grazing mammals would have been packs of wolves.

Wolves would have been found in the forests that covered most of northern Spain, France, Germany, Britain, the eastern European countries, and northern Russia. These forests made safe dwelling places for many animals, including several kinds of deer, bear, the wild boar, lynx, elk, and bison. Primitive man feared the forests. They were mysterious places full of large and dangerous animals. In fact, predatory animals rarely attack man unless first attacked themselves, but folklore told of dark and dreadful forests full of savage wolves and bears. When men eventually overcame their fears and entered the forests, they hunted these animals to near-extinction. Even today, when nearly all the large mammals of the region are rigorously protected, the wolf is still being exterminated.

In the modern world, too, man has reduced the wildlife around him in other, more devastating, ways. He has done this with pesticides, which kill not only the pests they are intended for, but also the animals, such as birds, that feed on the pests. Effluent from factories and domestic drainage has also taken its toll of wildlife.

The least spoiled areas of the Palaearctic region are the most inhospitable ones. The northern islands of Europe and Asia reaching up into the Arctic Circle have not changed in historic times. They still have their polar bears, seals, and walruses. Some still have the musk ox—the longest-haired living mammal. The bravery of these wild cattle has almost led to their extinction. They do not run from their traditional enemies, the bear and the wolf, or from their modern enemies— men with guns. Instead, the herd forms up into a tight circle with the calves in the middle and defends itself with hooves and horns.

Below, left: An Arctic fox in winter. In October the brown, summer coat is replaced by dense, white fur that serves to camouflage the fox as well as to protect it against the intense cold. Arctic foxes feed mainly on small mammals, especially lemmings.

Above: Grey, or timber, wolves in northern coniferous forest. Wolves feed mainly on carrion, and small animals such as mice, rabbits, and hares. Working in packs of up to two dozen animals, wolves occasionally kill a large animal such as a deer, first wearing it down to a state of exhaustion before they attack. Wolves travel up to 40 miles during a night's hunting and for this reason are difficult to keep in reserves.

Below: A snowy owl. Hunting by day as well as by night, the snowy owl swoops silently on to its prey of lemmings and other small mammals. In years when lemmings are scarce, snowy owls migrate southwards from the Arctic tundra in search of food.

Unfortunately these are no defence against bullets. Another animal that has not yet learned that man is an enemy is the polar bear, and it, too, is in danger of extinction.

South of these islands of ice and snow lie regions of frozen tundra. These are areas of treeless plains stretching in a circle around the North Pole. The sub-soil of the tundra is permanently frozen and little grows except creeping lichens and a few stunted bushes. Surprisingly, several animals choose to make their home on these barren flatlands the year round, and many migratory birds stop there to nest during the short, light summer.

Reindeer are the best-known mammals of the Palaearctic tundra. But there are few really wild reindeer left; most are domesticated animals bred to transport, feed, and clothe the men who herd them. The wolf and wolverine prey on these herds. The wolverine, which is also known as the glutton, has the habit of reserving food for its future consumption by marking it with a foul-smelling glandular secretion that repels competitors. The wolverine looks like a shaggy dog, but is, in fact, the largest member of the weasel family. Although mainly a scavenger, it is so fierce that even wolves and bears will leave a kill for it.

The smaller animals of the tundra blend with the wintry surroundings. The snowy owl is white all the year round, but the Arctic fox, ermine, and mountain hare moult their dark fur in the autumn and grow a white winter coat. The ptarmigan, a bird that lives on the tundra throughout the year, also turns from brown to white in winter. The small mammals—such as lemmings and voles—do not hibernate but remain active during the winter, living and searching for food under the snow layer, where the temperature is considerably higher than at the surface.

South of the tundra lies a vast tract of coniferous (evergreen) forest known by the Russian word *taiga*, meaning "swamp forest." This is the largest area of forest in the world. It stretches from the Scandinavian mountains in the west to the Pacific Ocean in the east. Underfoot it is often marshy.

The taiga is the home of the elk (a slightly smaller version of

25

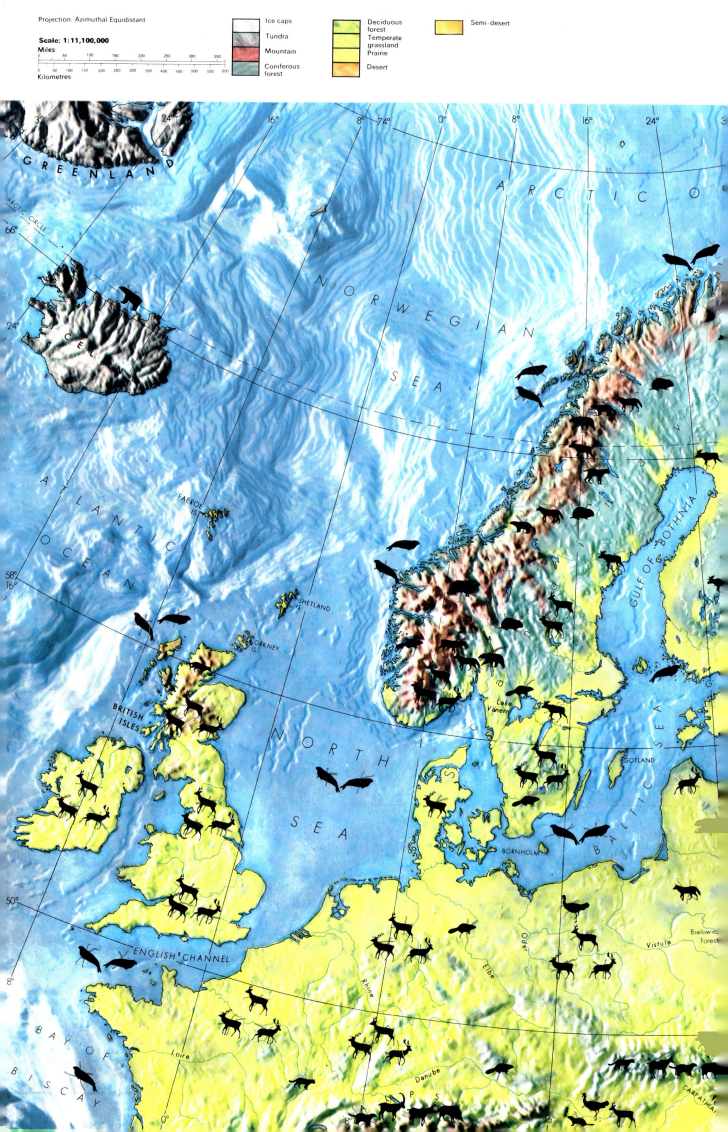

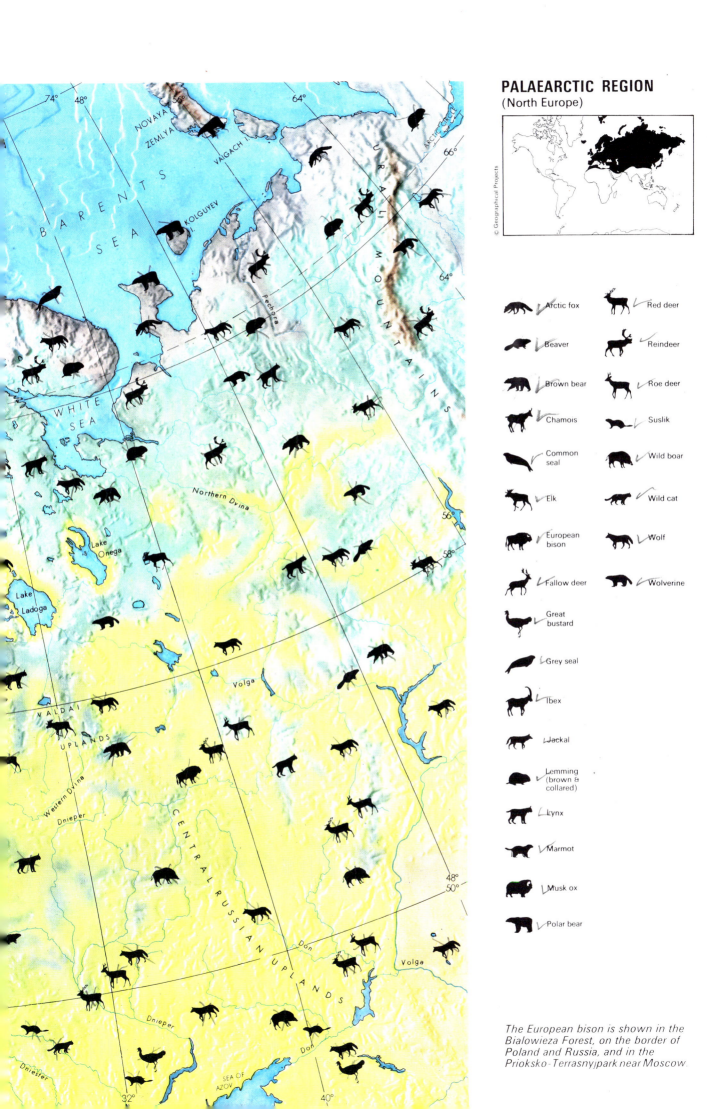

PALAEARCTIC REGION
(North Europe)

© Geographical Projects

Arctic fox

Beaver

Brown bear

Chamois

Common seal

Elk

European bison

Fallow deer

Great bustard

Grey seal

Ibex

Jackal

Lemming (brown & collared)

Lynx

Marmot

Musk ox

Polar bear

Red deer

Reindeer

Roe deer

Suslik

Wild boar

Wild cat

Wolf

Wolverine

The European bison is shown in the Bialowieza Forest, on the border of Poland and Russia, and in the Priosko-Terrasnyj park near Moscow.

the American moose)—the largest species of deer. Elks are tree-browsers, their height enabling them to reach leaves on the lower branches. They also wade into marshes or ponds to feed on water plants. Their broad hooves spread their weight as they move across the swampy or snowy ground in search of food. It was the elk's ability to move easily through swampy ground that, in the 1700's, led Catherine the Great of Russia to forbid the breaking in and riding of elks, because she thought they would be used by criminals to escape from authority.

Because the forests of the taiga are dense and difficult for man to penetrate, they have become a refuge for animals that are hunted elsewhere. The wolverine, the wolf, the brown bear, and the lynx are all found in the taiga—driven there because they are rival hunters of deer, and threaten man's domestic herds.

The sable, typical of the smaller animals of the coniferous forests, has been almost wiped out for a completely different reason. It has a very beautiful, warm pelt. The sable and its near relative, the marten, are tree-dwelling members of the weasel family and live alongside the squirrels and rodents on which they prey. Until about a century ago, the sable was a very common animal of the coniferous forests. In fact, there was such a glut of pelts in some parts of Russia that the sable began to be killed for meat, and the pelts discarded. But, as its numbers dwindled, the fur came to have a rarity value and the price paid for the pelts rose. Consequently, the animal was hunted with even more vigour and, except in some remote and mountainous parts of Russia, it disappeared. The sable was finally protected in 1913, and is now being bred successfully on fur farms to provide pelts for the fur industry.

The trees of the taiga provide protection, food, and nesting sites for many birds. Among them are woodpeckers, owls, and crossbills. In the swamps of the taiga and in the mixed forests of the south, where broad-leaved deciduous trees (those that lose their leaves during winter) grow among the conifers, many water-birds, including the curiously shaped spoonbill, make their homes. The water also attracts other animals. Beavers, for example, live on the outskirts of the taiga, where ash, willow, and birch—the bark of which is their favourite food—grow with pine and spruce. Frogs, toads, and newts are also plentiful.

There are many species of tailed amphibians in the Palae-arctic region, the most striking of which are the salamanders, especially the Japanese giant salamander, which can grow to a length of up to 5 feet. There is also the curious pinkish-white, blind olm, with its straw-like legs and ruff of pink gills, that lives in the caves of Yugoslavia.

South of the taiga and mixed forests is an area that many centuries ago was covered in dense deciduous forests. This is an area stretching across northern Spain, through Europe and Russia to the Altay Mountains of central Asia. This part of the region has a less extreme climate and has been invaded by man and altered beyond recognition. The trees have been cut down and the land has been cultivated. Only small patches of the forest remain, either where the country is mountainous or in

Below: An olm. Olms, which are primitive tailed amphibians, live in underground rivers, feeding on crustaceans and small fish. Immediately after they hatch, olm larvae are almost black and have well-developed eyes, but as they grow the colour disappears and the eyes sink into the skin. Full development takes 10 years, with the blind adults reaching a length of about 12 inches. Blindness and lack of colour are typical of cave-dwelling animals. Olms retain gills and a larval appearance throughout life.

Above: A group of red deer, consisting of a mature stag, a female, and several young. The stag loses its antlers in early spring, and a new set begins to grow about six weeks later, appearing first as skin-covered buds. The antlers reach full growth in summer. The skin, or velvet, covering them begins to dry up in August or September and is rubbed off, exposing the hard, and by then insensitive, bony core. The antlers become more elaborate each year, reaching the mature form in six years.

parks and nature reserves.

Paradoxically, the privileged landowners who hunted animals for sport on their private estates often saved these animals from extinction. Poaching by meat-hungry peasants was deterred by severe penalties. King Canute of England, who reigned between 1016 and 1035, imposed the death penalty on anyone caught stealing his deer. And European poachers of the Middle Ages risked having their hands cut off or their eyes put out. Those who owned the hunting rights made sure that enough animals were left to breed and produce more animals for them to hunt the next year. As a result, wild boar, red, roe, and fallow deer are still to be found in parkland in much of the Palaearctic region.

The largest surviving animal in the Palaearctic region, and one that was once widespread throughout the deciduous forests, is the European bison, or wisent. It is now found only in small numbers in the Bialowieza Forest on the borders of Poland and Russia and in a few nature reserves in Poland and Russia. The European bison, which is slightly larger than the American bison, was brought to the very brink of extinction but was saved by some of the European zoos. The last wild bison was shot in Bialowieza in 1921. Luckily, some of these animals had been presented by Tsar Alexander II to the Duke of Pless in 1865. Three survived, two bulls and one cow. It seemed as though the breed could not possibly continue. But, in 1918, the Director of Frankfurt Zoo formed The International Association for the Preservation of the Wisent and brought together all the bison still remaining in European zoos. By 1938 there were 100 of these huge animals and some were reintroduced into the Bialowieza Forest. Under protection there, they have become established as semi-wild cattle.

Some of the smaller animals such as voles, field mice, and many birds, have adapted themselves to the clearing of the land and now live in hedgerows rather than in real woodland. But even

Left: A young male Alpine ibex. Like other species of these wild goats, Alpine ibexes are expert climbers. They pick their way confidently among high mountain crags in search of the lichens and mountain plants on which they feed. In the Middle Ages the Alpine ibex was almost exterminated because of its supposed healing properties. The blood, for example, was claimed to cure kidney stones because the ibex lived on stony ground.

here they are threatened. Hedges and ditches are being replaced by barbed wire and other man-made fences that require less upkeep. Other animals find man a good neighbour. Swallows, starlings, house sparrows, pigeons, bats, rats, and house mice all make their homes in or on buildings. The harvest mouse builds its tiny spherical nest on the stalks of our cereal crops. Even the fox has learned to feed from garbage cans and dumping grounds and is now often found in town suburbs.

The mountains of Europe and Asia have a distinctive animal population, partly composed of animals that were once more widespread and have taken refuge in areas inaccessible to man. Thus the leopard and jackal have retreated to the Caucasus Mountains, and the lynx, brown bear, wild cat, wolf, and wild boar to one or other of the mountain ranges of Europe. More typical of mountains are the little marmots (ground squirrels) of the Alpine meadows, and many crag-loving animals such as eagles and vultures. Two nimble, goat-like animals, the ibex and the chamois, are also typical. The seemingly docile and lovable chamois is really a fierce little fighter that can disembowel a man with its sharp hooked horns.

The ibex is another animal that is recovering from near-extinction. It was a victim of mediaeval superstition, for people of the Middle Ages believed that nearly every part of the ibex was a remedy for some disease.

On the crags of the Atlas Mountains in North Africa live Barbary sheep, or aoudad, which are preyed on by the Barbary leopard. The leopard, however, is becoming increasingly rare because of the demand for its pelt.

The mountains of northern China are home to more unusual

animals. These include the snow leopard and the tiger, which also lives in the mountainous regions of Siberia and North Korea. Although the tiger is protected by the Russians and Koreans, it is still hunted in China, where its powdered bones are thought to give men the tiger's courage and fierceness. Another Chinese mountain animal is the giant panda, which lives in the dense bamboo forests on the borders of Tibet in the transitional zone between the Palaearctic and Oriental regions. In place of the chamois of European mountains the Chinese mountains have the goral, a kind of goat-antelope. A very similar animal, the serow, is found in the mountains of Japan and Taiwan.

The Altay Mountains, bordering the Gobi Desert in Mongolia, are among the most inhospitable areas in the world. On the open plains around the mountains live the remnants of two species of animals that were once widespread throughout the Asian plains. These are Przewalski's wild horse and the wild ass. Here too lives the wild, twin-humped Bactrian camel, an animal that once ranged over a wide area of the Gobi Desert. It is a sad reflection that these three species should be so near extinction while their domesticated descendants are so numerous.

South of the deciduous forest belt lie two different kinds of terrain. In the west lie the Mediterranean lands, and in the east are vast grasslands, the plains of Hungary, and the steppes of south-eastern Russia. These have been cultivated to some extent now, but there are still large tracts of rolling, dun-coloured land covered with grass and dotted with wild flowers.

Here, in surroundings similar to the prairies of America and the savannas of Africa, live many small rodents. They feed mostly on seeds and plant roots, and many hibernate to avoid the harsh winters of these areas. There are the jerboas and gerbils (mouse-like creatures with long hind legs for jumping), the susliks, or ground squirrels (cousins to the chipmunks and marmots), and the common hamster. The common hamster is larger and fiercer than the pet, or golden, hamster.

PALAEARCTIC REGION
(South Europe)

© Geographical Projects

Mountain	
Coniferous forest	
Deciduous forest	
Temperate grassland	
Prairie	
Mediterranean	
Savanna	
Desert	
Semi-desert Fertile lands	

Projection: Azimuthal Equidistant

Scale: 1:11,100,000

Miles
0   50   100   150   200   250   300   350

Kilometres
0   50   100   150   200   250   300   350   400   450   500   550   600

Addax   Brown bear

Barbary ape   Chamois

Barbary sheep   Common seal

Beaver   Crested porcupine

ATLANTIC OCEAN

BAY OF BISCAY

Loire

Rhine

Danube

L. Leman

A L P S

Po

LIGURIAN SEA

Camargue

Rhône

CORSICA

ADRIATIC

Douro

P Y R E N E E S

Ebro

Tagus

BALEARIC IS.

SARDINIA

M E D I T E R R A N E A N

TYRRHENIAN SEA

Guadalquivir

STR. OF GIBRALTAR

SICILY

G R E A T   A T L A S

S   a   h   a   r   a

TROPIC OF CANCER

42°
34°
26°

8°
0°
50°
8°
16°
0°
8°
16°

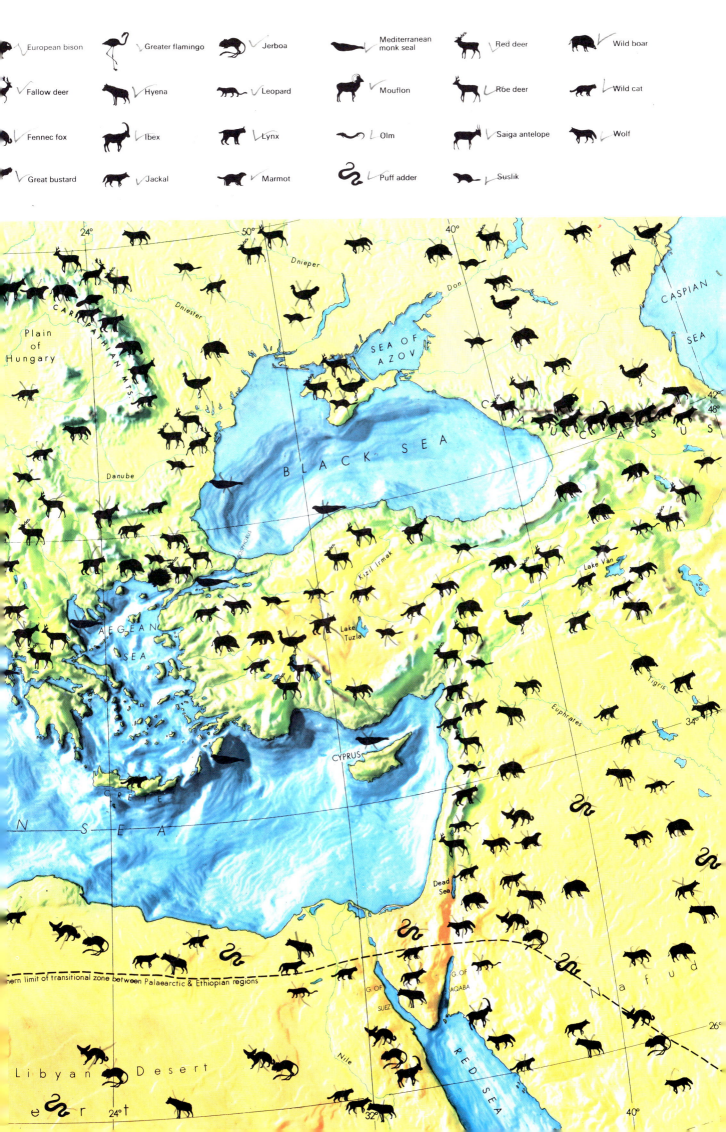

European bison   Greater flamingo   Jerboa   Mediterranean monk seal   Red deer   Wild boar

Fallow deer   Hyena   Leopard   Mouflon   Roe deer   Wild cat

Fennec fox   Ibex   Lynx   Olm   Saiga antelope   Wolf

Great bustard   Jackal   Marmot   Puff adder   Suslik

Left: A male saiga antelope. These
grassland animals, which stand about
2½ feet high at the shoulder, migrate
southwards in the winter to escape the
severe cold of the Russian steppes. The
intricate network of bones and mucous
membranes in the saiga's swollen nose
may serve either to warm inhaled air or
to filter out dust.

Both the hamster and the suslik collect food during the summer months and carry it to their burrows in cheek pouches. The store of food, which may weigh as much as 25 pounds, is used up during the winter when the animals go into a sort of semi-hibernation, waking up occasionally to feed. In the harshest part of the steppes, the suslik may spend as long as nine months below ground, dropping from a weight of 17 ounces to only 4 ounces. All their food collecting and breeding must be done in April, May, and June.

Other inhabitants of the steppes include the mole-rat, the marmot, and a variety of lizards and small snakes. Many birds, such as larks and pipits, also live here. The most spectacular bird of the area, however, is the great bustard, one of the largest flying birds. It is now very rare. The predators of the area, polecats, foxes, eagles and kites, feed mainly on the rodents.

Of the grazing animals that used to roam the steppes only the little saiga antelope is left. These animals are extremely sensitive to changes of temperature, and, by migrating away from the coldest areas, manage to survive the bitter winters of the Russian steppes. They are now strictly protected by the government of Russia, and even "farmed" in the sense that a limited number are shot each year to provide meat and hides. However, only recently has the saiga population recovered sufficiently for man to do this again. During the 1800's the saiga was almost wiped out because it was used to provide raw materials for Chinese medicines.

The animals of the Mediterranean lands arrived there from two sources. Some were pushed south by the glaciation during the Ice Age, and some moved north from Africa. At the beginning of the Christian Era the lion was common in the area, but now the only large predators are the lynx and the wild cat, and there are not many of these left. The ibex, once widespread,

is now found only in very restricted areas. It has bred with domestic goats, which are themselves probably descended from it. The wild sheep, or mouflon, and the fallow deer are also native to the area, although there are now more fallow deer in the northern European forests, where they have been introduced by man.

One of the Mediterranean animals that faces extinction is the monk seal. It originally lived all round the Mediterranean coast and the coast of north-west Africa. But now its beaches have been taken over by holidaymakers and, like all seals, it is persecuted by fishermen.

The more exotic members of the Mediterranean population are those that have spread from Africa. These include the Barbary apes of Gibraltar, and the crested porcupine found in parts of Italy and the Balkan peninsula. The mongoose and the cat-like genet of the Coto de Doñana in southern Spain are other examples. Many of the lizards of the area have come from Africa, as have the charming geckos with their clinging pad toes that enable them to run around on the ceiling. The Mediterranean chameleon, with its swivel eyes and long sticky tongue, is another fascinating animal that has spread from Africa.

Although much of the animal life of the Mediterranean is struggling hard for existence, there are two areas that reassure the naturalist. These are the Camargue—the delta of the river Rhône in France, and the Coto de Doñana, the delta of the Guadalquivir River in southern Spain. There are now nature reserves in both areas, the unspoilt vegetation providing a haven for wild animals and migratory birds. Some of these migratory birds arrive in the spring to nest, some in the autumn to over-winter, and many more use the areas as stopping-off places to feed and rest during their enormous migrations of 5,000 miles or more. The Camargue and the Coto de Doñana attract orni-

Left: A great bustard. These birds of central Asia are the largest of all flying land birds, i.e. those that normally alight on land. Male great bustards have reached a weight of over 46 pounds.

Above: Greater flamingos. These graceful birds gather in vast flocks consisting of many thousands of birds. Unlike lesser flamingos, which sieve microscopic plants from surface water, greater flamingos sweep up small snails and shrimps and also filter out organic matter from the mud.

Right: A Spanish lynx. Once common throughout most of Spain and Portugal, this animal is now restricted to certain controlled areas, the biggest population being, at most, 200 animals in the Coto de Doñana in southern Spain. The Spanish lynx is a race of the rare Mediterranean, or pardel, lynx, small populations of which may still survive in the Carpathian Mountains and in parts of Greece. Lynxes are known to prey on small mammals and birds, but naturalists have little reliable information on the Spanish lynx's way of life.

PALAEARCTIC REGION
(North Asia)

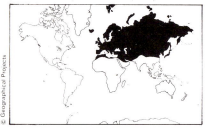

Arctic fox

Baikal seal

Beaver

Brown bear

Common seal

Elk

Fallow deer

Giant salamander

Great bustard

Grey seal

Himalayan black bear

Ibex

Japanese macaque monkey

Lemming (brown & collared)

Leopard

Lynx

Musk ox

Polar bear

Przewalski's wild horse

Red deer

Reindeer

Roe deer

Saiga antelope

Sika deer

Snow leopard

Suslik

Tiger

Wild ass

Wild Bactrian camel

Wild boar

Wild cat

Wolf

Wolverine

The wild ass symbol shown near the Aral Sea represents a group of Persian wild asses introduced into the island reserve of Barsakel'mes. The only surviving population of the Przewalski's wild horse is shown at the Tachin Shar Nuruu massif in south western Mongolia.

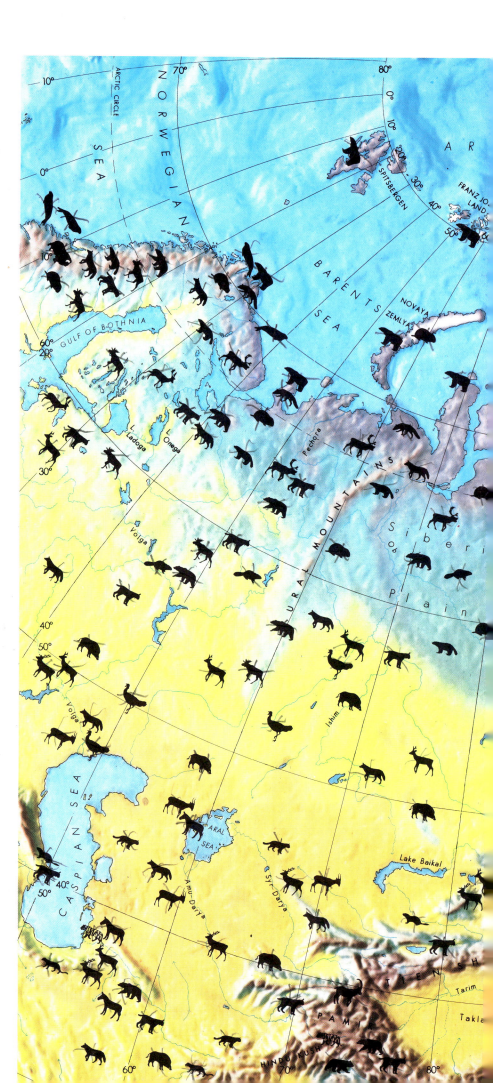

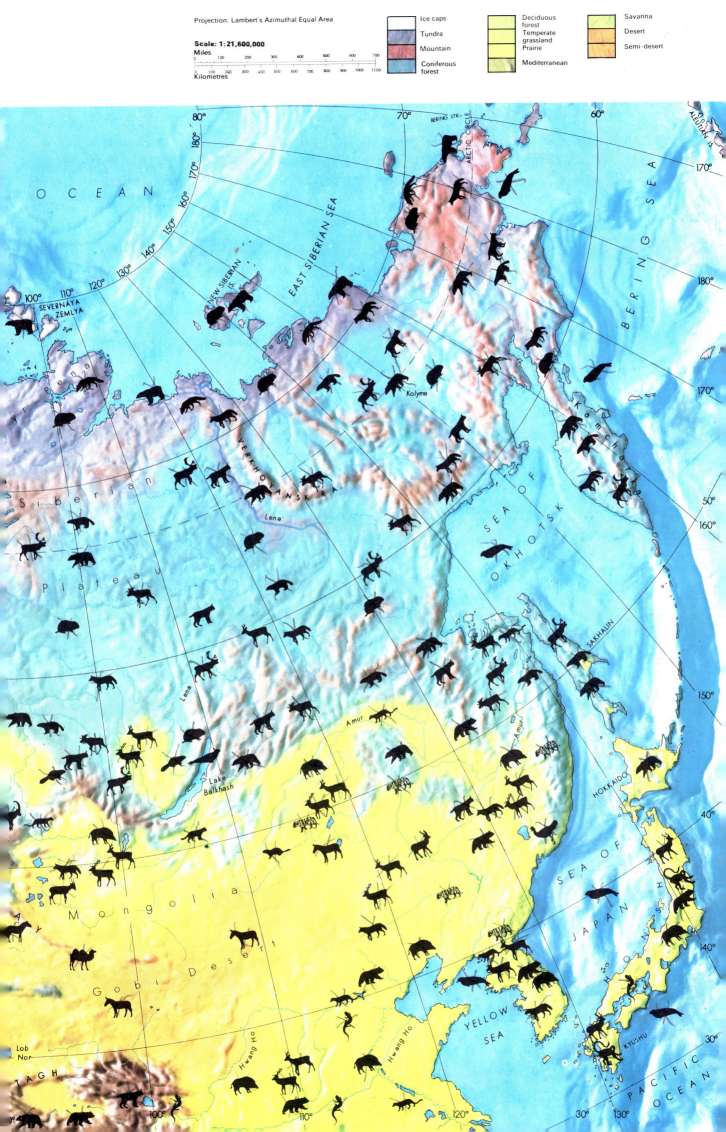

Animals of the Palaearctic Region

The chart lists the main groups of animals found in the Palaearctic region and shows if these animals occur in other regions.

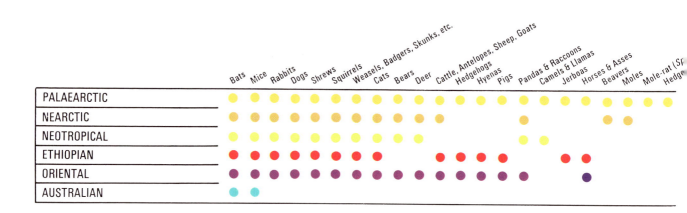

	Bats	Mice	Rabbits	Dogs	Shrews	Squirrels	Weasels, Badgers, Skunks, etc.	Cats	Bears	Deer	Cattle, Antelopes, Sheep, Goats	Hedgehogs	Hyenas	Pigs	Pandas & Raccoons	Camels & Llamas	Jerboas	Horses & Asses	Beavers	Moles	Mole-rat (Sp	Hedge
PALAEARCTIC	●	●	●	●	●	●	●	●	●	●	●	●	●	●	●	●	●	●	●	●	●	●
NEARCTIC	●	●	●	●	●	●	●	●	●	●	●				●				●	●		
NEOTROPICAL	●	●	●	●	●	●	●	●	●	●	●				●	●						
ETHIOPIAN	●	●	●	●	●	●	●	●	●		●	●	●	●			●	●				
ORIENTAL		●	●	●	●	●	●	●	●	●	●	●	●	●	●	●			●			
AUSTRALIAN	●	●																				

thologists who travel thousands of miles to see the ducks and wading birds, the flamingos, the eagles, egrets, vultures, and the brightly coloured hoopoes and bee-eaters.

Both areas are thinly populated by man, and half-wild domestic cattle and horses graze and splash through the marshy ground. Mammals that can no longer live in the rest of the Mediterranean region seek refuge in these reserves. The Coto de Doñana, for example, has red deer, wild boar, and the pardel, or Mediterranean, lynx, and even wolves occasionally visit the area.

At the southernmost edge of the Palaearctic region are areas of desert: in the west the Sahara and the deserts of Arabia; in the east the Gobi Desert of Mongolia. The animals of these deserts are few and specialized, as for most animals the combination of heat and lack of water is fatal. This is particularly so for mammals, because they depend on evaporation of water to keep cool.

Right: A fennec fox pounces on a desert jerboa. The fennec is the smallest of all foxes, with a head and body length of about 15 inches. It eats a variety of foods, including plants, insects, lizards, birds, and small rodents. Like the jerboa, the fennec burrows into the sand during the day and becomes active only during the cool desert night. The jerboa moves along the ground in a series of leaps searching for the roots, seeds, and grass on which it feeds. When standing, the jerboa uses its long tail as an extra support.

Most of the small mammals that live in these inhospitable areas are nocturnal and so avoid the heat of the day. Some give up the struggle to survive during the hot summer months and go into *aestivation*—the summer counterpart of hibernation. The successful rodents of the Sahara, such as the jerboa and gerbil, have developed ingenious ways of conserving vital supplies of water. Like the kangaroo rats of North America, these animals can survive without drinking any water at all. They can derive enough water from the dry seeds they eat, and conserve it by producing a very concentrated urine. Also, they have no sweat glands and so lose little water by evaporation.

During the Sahara night, these animals are preyed upon by jackals, hyenas, and by the beautiful little fennec fox with its disproportionately large ears. These extraordinary features act as radiators of heat and also enable the fox to locate its prey accurately.

The large mammals of these desert regions—the wild asses and camels of the Gobi (a desert hot in summer and cold in winter) and the domestic camels of the Sahara—cannot avoid the heat and thirst. The camel does not, as is generally believed, carry a store of water in its hump or stomach. The hump is a store of fat and serves to nourish the animal when food is scarce. The camel's main adaptation to heat and lack of water is its ability to maintain the water content of its blood, but lose water instead from the tissues. (In most mammals the blood thickens and death soon follows.) The wild ass also can survive the reduction in body weight that occurs during this process. When water is available, both these animals can drink and absorb enough water to make up the loss. This may involve a camel's drinking as much as 27 gallons of water at one time. There are also two antelopes that survive in these harsh surroundings—the addax antelope of the Sahara and the Arabian oryx, which is now extremely rare.

Reptiles, particularly snakes and lizards, fare rather better in these deserts. They have impermeable skins that prevent water loss to the air, and they also produce a semi-solid, highly concentrated urine. The spiny-tailed lizard, which occurs throughout the Palaearctic deserts, builds up a food store of fat in its tail. The most common snakes of these deserts are vipers. There is the horned viper of Africa and south-western Asia, and the highly venomous puff adder of Africa.

Being cold-blooded, the desert reptiles regulate their body temperature by moving in and out of the sun and shade. Some lizards also change colour to regulate their absorption of the sun's rays. Only by a combination of behavioural and physical adaptions can the reptiles, along with the other desert animals, survive in what is one of the world's most hostile environments.

3 The Nearctic Region

Greenland, Canada, the United States (including Alaska and the Aleutian Islands, but not Hawaii), and the desert and semi-desert area of Mexico as far south as the Tropic of Cancer.

Man's impact on the wildlife of the Nearctic region has been recent and dramatic. The disruption that took place over a thousand years in the Palaearctic region has been telescoped in the Nearctic region into the few centuries after the settling of the white man in the early 1600's. The original inhabitants of the Americas—the Eskimos and the Indians, or Amerinds—were so few in number, and their weapons so unsophisticated, that they had little effect on the animal world. The white man, however, started an unprecedented slaughter. But while many people displayed an almost maniacal desire to kill the wildlife, others took action to preserve it. As a result, national parks were set up. The first one was the Yellowstone National Park, founded in 1872—thirty-seven years before the first park was formed in Europe.

The Nearctic region has very much the same climatic and vegetational zones as the Palaearctic region. Not surprisingly, the animal population is almost identical. In fact some naturalists join the two areas together under the name of the Holarctic region. The chart on page 51 shows that the Nearctic region also has animals in common with the Neotropical region (Central and South America). Some of these, the opossums and armadillos for example, have spread into North America from South America.

There are several unique animals in North America, such as the wild turkey and the pronghorn, but there are also some strange gaps in the animal population. There are no camels or wild horses, even though these animals originated in prehistoric times in North America. They migrated into South America, Asia, and Africa, and then died out in North America. (Horses also died out in South America.) The horses used by the

Brown bears catching fish in a mountain river. Most brown bears' varied diet also includes berries, wild bees' grubs, small mammals, and sometimes larger mammals such as young deer.

NEARCTIC REGION
(Canada & Greenland)

© Geographical Projects

Ice caps	
Tundra	
Mountain	
Coniferous forest	
Deciduous forest	
Temperate grassland	
Prairie	

Projection: Lambert's Equal Area

Scale: 1:13,500,000

Miles
100   200   300   400

100   200   300   400   500   600   700
Kilometres

Arctic fox
Black bear
Beaver
Bobcat
Bighorn sheep
Brown bear
Bison
Caribou

BEAUFORT SEA

BANKS ISLAND

VISCOUNT MELVILLE SD.

DEVON

VICTORIA ISLAND

GULF OF

BROOKS RANGE

ARCTIC CIRCLE

ALASKA

Yukon

MACKENZIE MTS.

Mackenzie

Great Bear Lake

Great Slave Lake

Lake Athabasca

ALEXANDER ARCHO.

Peace

Athabasca

Nelson

QUEEN CHARLOTTE IS.

Saskatchewan

Saskatchewan

Lake Winnipeg

VANCOUVER ISLAND

JUAN DE FUCA STR.

Fraser

Columbia

South Saskatchewan

Lake Manitoba

PACIFIC OCEAN

Columbia

Missouri

Yellowstone

70°   160°   150°   140°   130°   120°   110°   100°   90°

150°   60°   140°   50°   130°   120°   110°   100°

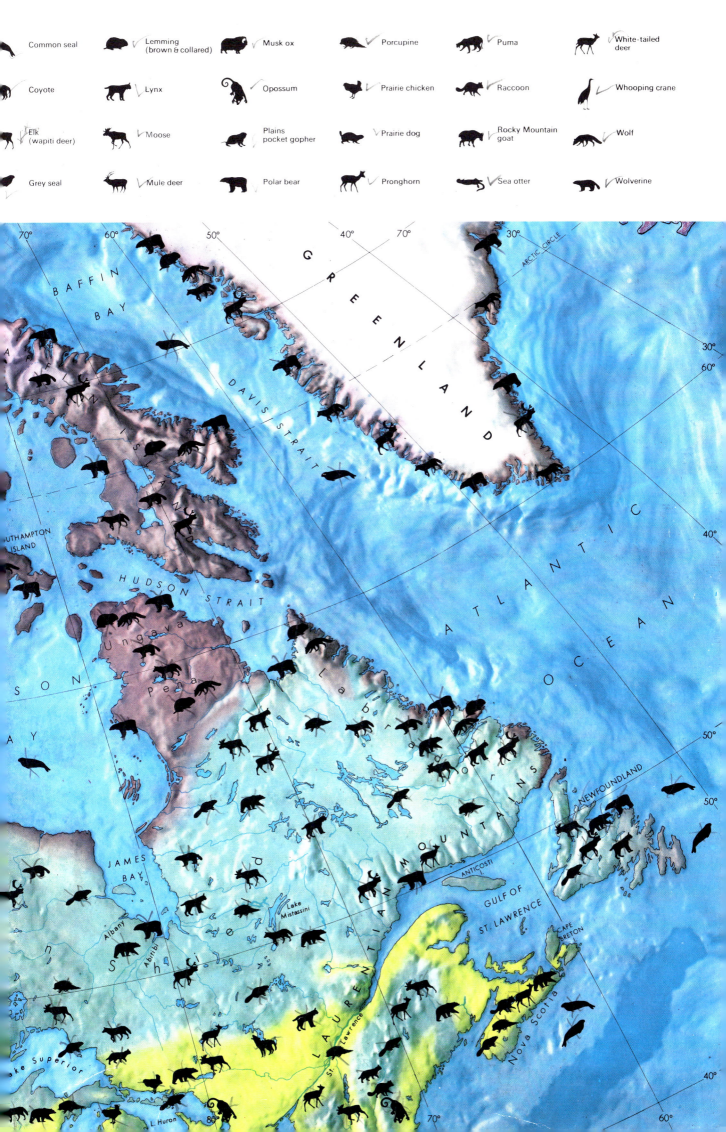

Common seal

Lemming (brown & collared)

Musk ox

Porcupine

Puma

White-tailed deer

Coyote

Lynx

Opossum

Prairie chicken

Raccoon

Whooping crane

Elk (wapiti deer)

Moose

Plains pocket gopher

Prairie dog

Rocky Mountain goat

Wolf

Grey seal

Mule deer

Polar bear

Pronghorn

Sea otter

Wolverine

70° 60° 50° 40° 70° 30°

G R E E N L A N D

BAFFIN BAY

DAVIS STRAIT

ARCTIC CIRCLE

30°
60°

40°

ATLANTIC OCEAN

BAFFIN ISLAND

SOUTHAMPTON ISLAND

HUDSON STRAIT

Ungava Peninsula

H U D S O N   B A Y

Labrador

50°
50°

NEWFOUNDLAND

JAMES BAY

Lake Mistassini

Albany

Abitibi

ANTICOSTI I.

GULF OF ST. LAWRENCE

CAPE BRETON

L A U R E N T I A N   M O U N T A I N S

St. Lawrence

Nova Scotia

Lake Superior

L. Huron

40°

60°

80°

70°

Above: A beaver. Beavers of North American rivers build dams of branches, stones, and mud to form ponds in which they live and store their food. Powerful jaws and sharp incisor teeth enable beavers to fell trees and strip off the bark, which they eat.

Amerinds were the wild descendants of those brought over by the Spanish conquistadors in the early 1500's.

As in the Palaearctic region, the zone least changed by man is the ice and tundra region in the north. This zone includes Greenland, Baffin Island and the other Canadian islands, and the northernmost parts of the Canadian mainland and Alaska.

Around the North Pole the Arctic Ocean is frozen into a great shifting, cracking ice-cap. This area is the home of polar bears, whales, and seals. In the snows on the islands and mainland live Arctic foxes, lemmings, weasels, ptarmigans, snowy owls, Arctic terns, and gulls. Musk oxen are found in the cold regions of Greenland and Canada. Here, too, lives their predator, the wolf. Caribou, which are closely related to reindeer, spend the spring and summer in the north, but migrate in the autumn to the coniferous forests many hundreds of miles to the south.

Off the coast of Alaska live three interesting marine mammals: the walrus, the northern fur seal, and the sea otter. The last two animals were almost exterminated for their beautiful coats. They are now protected to some extent and are increasing in number. The sea otter is well adapted for life in water, its hind limbs being broad and flipper-like. In fact, it can move only clumsily on land. It feeds and gives birth to its young about half a mile offshore in thick beds of seaweed called kelp. It also anchors itself to these ribbon-like plants while sleeping. Sea otters spend much of their time floating on their backs with their babies or their food balanced on their chests. They will sometimes balance stones in this way on which they crack open shellfish.

South of the tundra lies a great belt of spruce forest, which is very marshy underfoot and scattered with lakes. Here live moose, white-tailed deer, and, during the winter months, caribou. The main predators of the area are eagles, hawks, wolves, wolverines, lynxes, martens, and fishers (like the wolverine the last two are members of the weasel family).

In Alaska, and the islands around its western coast, live the black and brown bears that were once widespread throughout the forests. These bears are misnamed because individuals of each species can vary in colour from pale grey through brown to black. Black bears are slightly smaller and more rounded in

Above: A raccoon. These solitary
animals live near water, in hollow trees
or rock dens. They eat many kinds of
animals and plants, but fish, frogs, and
other aquatic animals are their favourite
food. Raccoons are good swimmers and
climbers and are most active at night.

shape than brown bears, and less dangerous to man. They do not
shun man's company, as do brown bears, and in many wildlife
parks they have become persistent and insolent beggars.

The lakes of the area are home to countless water-birds,
including gulls, ducks, and swans. The commonly domesticated
white mute swan was introduced from Eurasia, but the whistling
and trumpeter swans are native to North America. The very
rare whooping crane, noted for its loud whoop-like call, breeds
in river swamps south of the Great Slave Lake in northern
Canada.

At the borders of the coniferous forests and the deciduous
forests to the south lives the beaver. Originally, this animal
had a very wide distribution in the Nearctic region. But from the
1600's onwards, fur-trapping was the chief industry of the region
and beaver pelts were the most important export. Other animals,
such as the ermine, otter, marten, fisher, wolverine, grey wolf,
and black bear, also provided fur. The Russians in the west and
the French, English, and Dutch in the east settled the coastal
regions and then ventured inland to trade in pelts with the
Indians. The British formed the Hudson's Bay Company to
further their fur-trading interests. The company would trade a
gun to the Indian trappers in exchange for a pile of beaver furs
as tall as the gun. And by the time the beaver hat went out of
fashion at the beginning of the 1800's there were not many
beavers left.

The area of deciduous woods south of the Great Lakes has
been changed immensely by farming and industrialization.
Some of the animals, such as raccoons, red foxes, opossums,
chipmunks, owls, and grey squirrels, have come to terms with
urbanization and have been recorded in or near cities. Other
animals have taken to the hills. White-tailed deer, black bears,
porcupines, and coyotes are all found in the Appalachians.
Many of the less easily seen species, such as the strange star-
nosed mole, snakes, turtles, newts, and salamanders, also live in
these mountains.

Tailed amphibia are typical of the Nearctic and Palaearctic
regions. Like the Palaearctic, the Nearctic boasts a giant salamander, known as the hellbender. It grows to a length of 11 to
30 inches and lives in the tributaries of the Mississippi. The wild

Left: A whooping crane. This magnificent 4-foot-tall bird is extremely rare, there being
only about fifty individuals surviving, including several bred in captivity. In the wild the
whooping crane breeds in river swamps south of the Great Slave Lake, Canada, and
winters in the Aransas Refuge on the Texas coast.

D·N

turkey is also typical of the eastern United States, but it has largely died out now and has been replaced by its domestic descendant.

South of the forest regions lie the vast prairies—a great expanse of land running across the continent from the northwest to the south-east. Most of this area is now used for raising stock and growing wheat, but it was once one enormous grassy plain blackened by herds of American bison, or buffalo.

In the early 1800's there were an estimated 60 million bison. By 1883 virtually all were dead. As the West was opened up, the settlers began to kill bison for food, but it was the laying of the railroads that sounded their death knell. Professional hunters, such as Buffalo Bill Cody, were employed to feed the gangs of labourers building the railroads. Buffalo Bill claimed to have killed 4,280 bison in 1869 alone for the Union Pacific Railroad.

The completion of the railroads made available the eastern market for meat and hides, and a mad slaughter began. When conservationists began to agitate for legislation to curtail the massacre of bison it was bitterly opposed in Washington. The argument behind this opposition was that once the bison were gone the warring plains Indians, such as the Sioux, who depended on these animals for food and shelter, would be defeated. The last large herd was exterminated in North Dakota in 1883. Only two small herds survived, one in Canada and one in Yellowstone Park. Bison are now vigorously protected against both man and disease, and small herds have been established in several national parks.

Pronghorn antelopes, mule deer, and elk also grazed the prairies in large herds during the 1700's. (Elk, known, too, as wapiti deer, are closely related to European red deer.) The pronghorn, of which about 350,000 remain of an original 40 million, is unique to the Nearctic region and the only antelope-like animal found in North America. It is not a true antelope—these are found only in Africa and Asia—but a halfway stage between an antelope and a deer. It has a bony core to its horns like an antelope, but sheds the horny outer covering like a deer. These little animals are extremely swift and can outrun a horse. But their habit of investigating anything unusual, such as a rag tied to a stick or a pair of moccasins sticking out of sagebrush, made them easy prey to hunters in the past. If alarmed they raise white hairs on their rumps as a signal to the rest of the herd.

The smaller animals of the prairies include the prairie chicken, the chipmunk, and the prairie dog, a ground squirrel that lives in vast and well-ordered underground towns. Another small burrowing animal typical of the area is the pocket gopher. The name refers to the cheek pouches, or pockets, in which it carries food to a storage place. All these animals are preyed upon by coyotes, American badgers, skunks, and prairie falcons.

Prairie dogs and pocket gophers play an important part in the upkeep of the prairie. By grazing the tops of plants and cropping long roots they weed the prairie and check the spread of scrub. Where these animals have been exterminated, thistles, then shrubs, and finally trees, especially the almost indestructible mesquite, have invaded the prairies from the scrub belts.

Above: Prairie dogs at the entrances to their burrows. These ground squirrels derive their name from the sharp, dog-like bark they utter when threatened. Below: A pronghorn antelope. These swift, sharp-eyed, inquisitive animals are unique to North America, where they roam the grassland and desert areas feeding on shrubs and grass.

Above: An American bison. These massive animals may weigh over 2,500 pounds and stand over 6 feet high at the shoulder. During the mating season (July to September) the breeding bulls fight among themselves for possession of the cows. These fights are usually shows of force, but occasionally bulls wound each other with their sharp, up-curving horns.

On the western side of the Nearctic region are vast mountain ranges, the Rocky Mountains in the north and the Sierra Nevada and Sierra Madre Occidental to the south. As in the Palaearctic region, the mountains provide a home for animals that once had a wider distribution but now survive only in areas not densely populated by man. This applies to the black bears and elk found in the mountains and also to bighorn sheep, which have been hunted for their magnificent curved horns. The long-haired Rocky Mountain goat, which is related to the chamois, is a native of the mountains, however, and lives on the highest and most inaccessible peaks of the Rockies. It is a splendid climber, its short strong legs with small hooves enabling it to leap safely from one narrow mountain ledge to another.

The main predators of the mountains are the puma, or mountain lion, and the lynx. South of the Canadian border the lynx is replaced by the bobcat.

Many small gnawing animals live in the western part of the Nearctic region, including the marmot, the pika, or whistling hare, which looks like a short-eared rabbit, and the mountain beaver, or sewellel. The sewellel is a thick-set, burrowing animal that lives near forest streams. It is unique to the Nearctic region and found only in the extreme west of the continent.

Bison symbols plotted on the maps in this chapter represent herds kept in the following places (from north to south): Upper Nyarling River (Mackenzie); Wood Buffalo NP (Mackenzie/Alberta); Elk Island NP (Alberta); Waterton Lakes NP (Alberta); National Bison Range (Mont.); Theodore Roosevelt National Memorial Park (N. Dak.); Yellowstone NP (Wyo.); Finney County State Game Refuge (Kans.); Wichita Mountains NWR (Okla.).

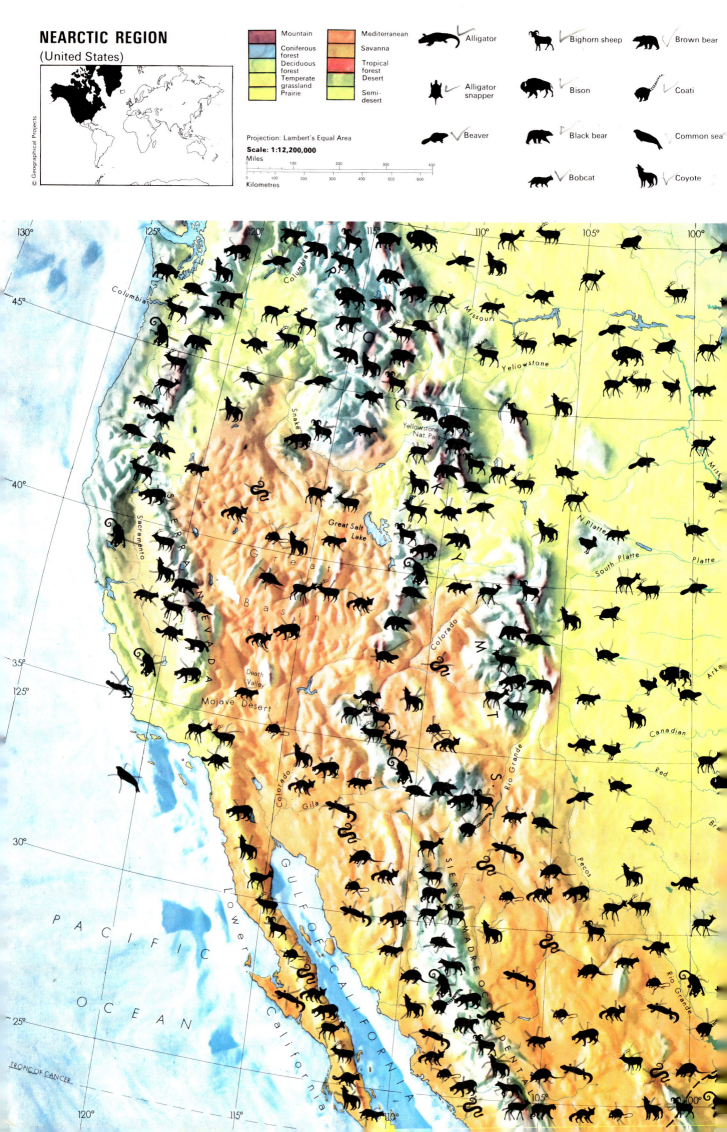

NEARCTIC REGION
(United States)

© Geographical Projects

| Mountain |
| Coniferous forest |
| Deciduous forest |
| Temperate grassland |
| Prairie |
| Mediterranean |
| Savanna |
| Tropical forest |
| Desert |
| Semi-desert |

Projection: Lambert's Equal Area
Scale: 1:12,200,000
Miles
0   100   200   300   400
Kilometres
0 100 200 300 400 500 600

Alligator

Alligator snapper

Beaver

Bobcat

Bighorn sheep

Bison

Black bear

Brown bear

Coati

Common sea

Coyote

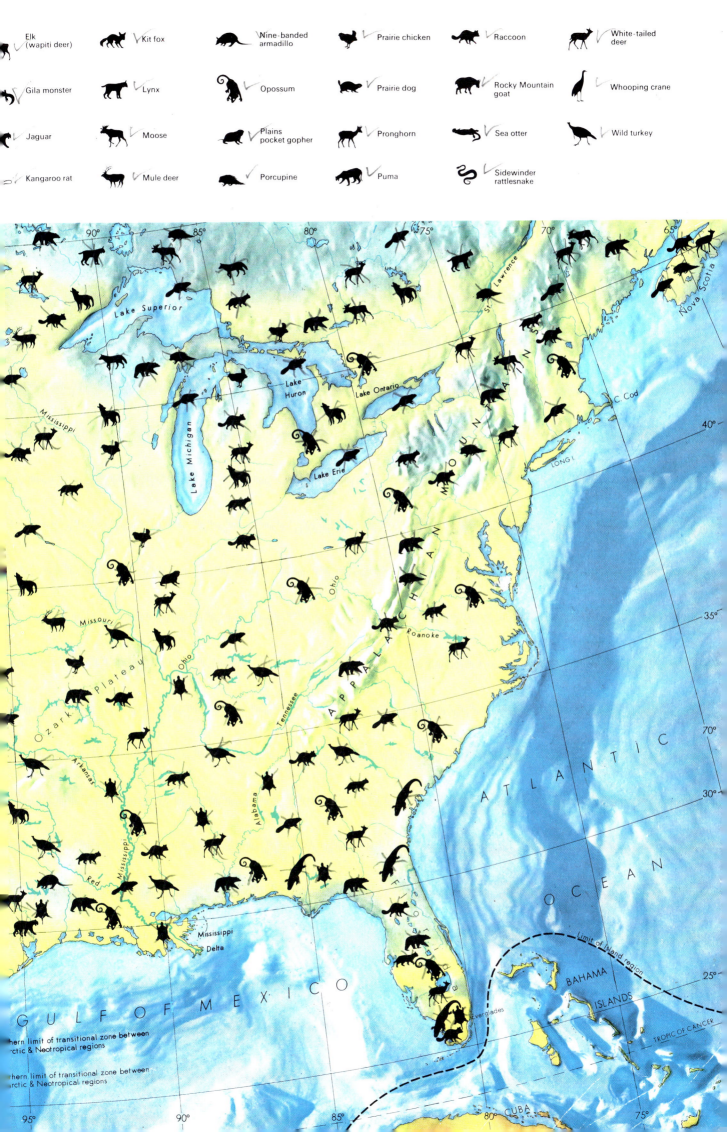

Elk (wapiti deer)
Kit fox
Nine-banded armadillo
Prairie chicken
Raccoon
White-tailed deer

Gila monster
Lynx
Opossum
Prairie dog
Rocky Mountain goat
Whooping crane

Jaguar
Moose
Plains pocket gopher
Pronghorn
Sea otter
Wild turkey

Kangaroo rat
Mule deer
Porcupine
Puma
Sidewinder rattlesnake

Lake Superior
Lake Michigan
Lake Huron
Lake Erie
Lake Ontario
St. Lawrence
Nova Scotia
C. Cod
LONG I.
Mississippi
Missouri
Ozark Plateau
Ohio
Arkansas
Red
Mississippi
Ohio
Tennessee
Alabama
APPALACHIAN MOUNTAINS
Roanoke
Florida
Mississippi Delta
Everglades
GULF OF MEXICO
ATLANTIC OCEAN
BAHAMA ISLANDS
Limit of Island region
TROPIC OF CANCER
CUBA

_ern limit of transitional zone between _ctic & Neotropical regions

_hern limit of transitional zone between _rctic & Neotropical regions

The deserts of the south-western American states and Mexico are, like all deserts, populated by very specialized animals. These include kangaroo rats, pack rats, pocket mice, and long-eared jackrabbits. All are able to withstand the heat and drought and are seemingly oblivious to the sharp cactus spines. They are preyed upon by the large-eared kit fox and the shy cacomistle, a relative of the raccoon, but far more beautiful with large eyes and a black-and-white ringed tail.

Several birds use the cactuses for nesting and profit by the defence offered by the spines. Woodpeckers hollow out the cactus stems and their abandoned holes are then used by the small elf, or cactus, owl. Cactus wrens and even hawks also make their homes in these prickly plants.

Large numbers of reptiles live in the desert, leaving the shade to sun themselves when the worst heat of the day is over. Among the many snakes living in these barren areas are the poisonous sidewinder rattlesnake and the equally deadly Arizona coral snake. The Mexican desert is unique in possessing the only poisonous lizards in the world, the slow-moving gila monster and the beaded lizard. The horned toad is also typical of the area, but, despite its name and its ferocious appearance, it is a harmless reptile. Not so harmless are the scorpions and the black widow and tarantula spiders of these deserts.

Less specialized animals also penetrate into the desert regions. From the north come coyotes, skunks, bobcats, and white-tailed deer; from the south, ocelots and jaguarundis (both wild cats), and the raccoon-like coatis.

Below: A desert scene with a sidewinder rattlesnake and a gila monster. Both of these desert reptiles are poisonous and track down prey by flicking their tongues over the ground and transferring tell-tale chemicals to the sensitive Jacobson's organ in the roof of their mouths. The rattlesnake also uses the heat-sensitive organs on its head to detect the presence of warm-blooded animals.

Above: An alligator snapper. This fresh-water turtle has a worm-like lure on its tongue with which it attracts fish and young water birds within range of its powerful jaws. Below: An American alligator. These sluggish animals, which may reach 19 feet in length, feed mainly on fish and small mammals. They incubate their eggs in heaps of rotting vegetation on the river bank.

In the valleys between the sierras of Mexico a tropical vegetation supports a number of exotic animals. Parrots and hummingbirds are common, and coatis and even jaguars can be found. In the lakes of Mexico lives the axolotl, a strange kind of salamander. It lives its entire life, and even breeds, while still in its larval stage. It has a pink or black flattened body and a frill of gills around its neck.

Two other areas of the Nearctic region have an almost tropical vegetation. These are the delta of the Mississippi River and the Everglades of the Florida Peninsula. Both these areas have lagoons and swamps that are overgrown with palms, mangroves, and cypresses. In these surroundings live many aquatic animals. Muskrats build their lodges here, and raccoons, otters, and nutria (coypus) live in the banks of the waterways.

In the brackish estuaries of the Everglades lives an extraordinary mammal called the manatee. It is a strange, seal-like animal equipped with front flippers but no back limbs. Like its marine relation, the dugong of east African, Asian, and Australian waters, it feeds on aquatic plants and never leaves the water.

Large numbers of birds live in these two areas. Wading birds such as herons, egrets, and flamingos are common. The galinule runs over the lily pads on its extremely long weight-spreading toes, and the osprey swoops to catch fish in its talons. Reptiles, too, are plentiful. There are tree and water snakes, and also many turtles, including the alligator snapper with its powerful jaws and vicious bite. It has a worm-like lure on its tongue with which it tempts unwary fish. Alligators glide silently through the dark waters and occasionally the very rare American crocodile can be seen.

Animals of the Nearctic Region

The chart lists the main groups of animals found in the Nearctic region and shows if these animals occur in other regions.

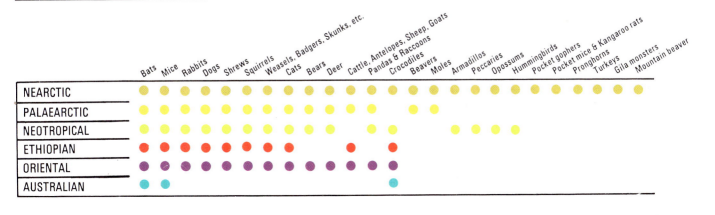

	Bats	Mice	Rabbits	Dogs	Shrews	Squirrels	Weasels, Badgers, Skunks, etc.	Cats	Bears	Deer	Cattle, Antelopes, Sheep, Goats	Pandas & Raccoons	Crocodiles	Beavers	Moles	Armadillos	Peccaries	Opossums	Hummingbirds	Pocket gophers	Pocket mice & Kangaroo rats	Pronghorns	Turkeys	Gila monsters	Mountain beaver
NEARCTIC	●	●	●	●	●	●	●	●	●	●	●	●	●	●	●	●	●	●	●	●	●	●	●	●	●
PALAEARCTIC	●	●	●	●	●	●	●	●	●	●	●			●	●										
NEOTROPICAL	●	●	●	●	●	●	●	●	●	●	●		●			●	●	●	●						
ETHIOPIAN	●	●	●	●	●	●	●	●			●		●												
ORIENTAL	●	●	●	●	●	●	●	●	●	●	●	●	●												
AUSTRALIAN	●	●											●												

4
The Neotropical Region

The tropical part of Mexico, Central and South America, including Trinidad and Tobago (but not the other islands of the West Indies), and the Falkland Islands.

South America, the largest part of the Neotropical region, is always thought of as an exciting continent. It is a land of mystery—of El Dorado, the mythical golden city that the conquistadors searched for so diligently in the 1500's. It is also a land of extreme geographical features. South America has the world's largest tropical rain forest, and the longest river. In the forest there are still places unvisited by white men, and perhaps human tribes still undiscovered. In keeping with these surroundings the animals are as strange as anything in this continent.

The Neotropical region is the original home of the opossums—the only pouched (marsupial) mammals to be found outside the Australian region. The toothless (edentate) order of mammals, including sloths, anteaters, and armadillos, is unique to the region, although one species, the nine-banded armadillo, has spread northwards into the southern United States. Even the rodents, monkeys, and bats of the Neotropical region are different from their relations in other parts of the world. The birds, too, are exciting and unusual; so much so that South America has been called the "bird continent." The region is also the home of enormous constricting snakes, ant-eating toads, electric eels, lungfish, and the carnivorous piranha fish.

The present-day isthmus between the North and South American continents has existed for only a few million years, although possibly the two regions were joined for a short time at an earlier date. For many millions of years, therefore, South America has been separated from other land masses and the animals have developed in isolation. Thus the Neotropical region has more unique animals than any other region. The chart on page 67 shows, however, that some Neotropical animals are shared with unlikely and far-distant regions. Tapirs, for

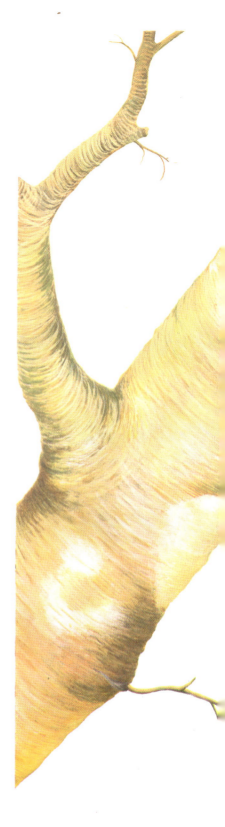

A two-toed sloth. Like its relative, the three-toed sloth, this forest-dwelling mammal spends most of its life hanging upside down. Although slow in most of its activities, the sloth can strike fairly quickly with its arms when threatened, inflicting severe wounds with the large, sharp claws.

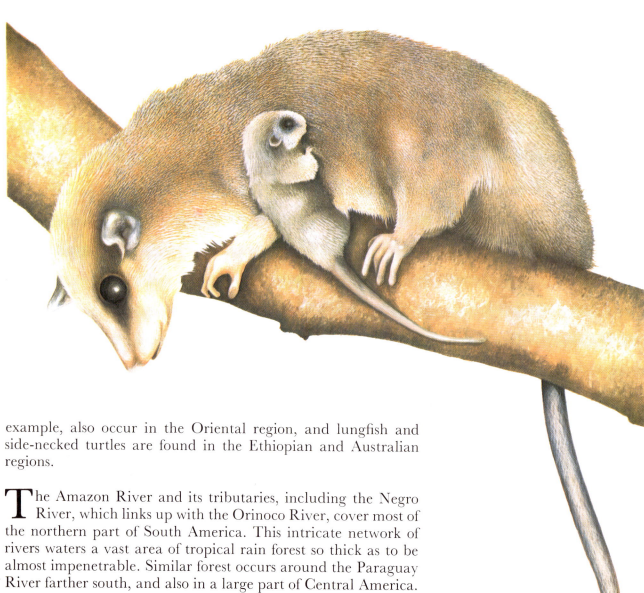

example, also occur in the Oriental region, and lungfish and side-necked turtles are found in the Ethiopian and Australian regions.

The Amazon River and its tributaries, including the Negro River, which links up with the Orinoco River, cover most of the northern part of South America. This intricate network of rivers waters a vast area of tropical rain forest so thick as to be almost impenetrable. Similar forest occurs around the Paraguay River farther south, and also in a large part of Central America.

The rivers are the highways of the forest. Until the development of air travel they provided the only means of penetrating the area. The sheer difficulty of cutting through the dense undergrowth has been one of the main factors limiting man's spread into the forest. The biting insects, many of them capable of transmitting dangerous diseases, have also kept man away. Thus the forest wildlife has remained largely unaffected by man. This means that little is known about some of the most inaccessible creatures and that there are few laws protecting the wildlife of the area.

The insects may be a deterrent to man, but they form the food supply of many of the forest animals. The opossums, for example, feed on insects, although the woolly opossum also eats fruit and seeds. In common with other pouched animals, the opossum gives birth to babies at an early stage of development. The babies crawl into the pouch—a mere flap of skin in this case—and attach themselves to the mother's teats. Once fully developed, the babies ride on their mother's back.

When the Virginian opossum (which lives in North, Central, and South America) is attacked it shams death or "plays possum." It lies still, closes its eyes and lolls out its tongue. In

Above: A mouse opossum with young. These small, usually tree-living animals feed mainly on insects, their prey including giant grasshoppers that can be almost as big as the opossums themselves. Mouse opossums rest during the day, usually in nests built from leaves and twigs. They are expert climbers, using their prehensile tails for extra grip when climbing vertically.

Map overleaf: The animals shown in the Nearctic region are those that occur in both the Neotropical and Nearctic regions.

this state it will allow itself to be mauled about until the predator loses interest—or eats it.

The forest-dwelling anteaters, the tamandua and the dwarf anteater, also eat insects. They break open termite nests with their strong front claws and explore the galleries with their long sticky tongues. The termites do not have time to sting as the anteater draws its insect-covered tongue into its mouth. These anteaters are tree-living animals with prehensile (grasping) tails. Like the giant anteater, which is found on the forest floor and also on the grasslands, they have no teeth, being members of the edentate order.

A forest-dwelling edentate that does have simple, peg-like teeth, however, is the giant armadillo. Like the anteater, it uses its enormous front claws to break open termite colonies on the forest floor.

The sloths, also edentates, have back teeth that they use to grind up their diet of leaves and fruit. Sloths are highly specialized for their tree-top life. They hang upside down on large hook-like claws and remain motionless for days, depending on their immobility and their greenish tinged fur to hide them from predators. The fur is parted along the belly and hangs downward towards the back so that the tropical rains run off it. Sloths are almost helpless on the ground and can barely move along.

The tropical forest is also the home of many insect- and fruit-eating bats. The most famous, or infamous, of these is the vampire bat, which feeds on mammalian blood. To anyone brought up on horror movies these are disillusioningly small, being only about three inches long with a wing span of about six inches. They have well-developed incisor teeth with which they slice off a thin layer of skin from their prey. A substance in the saliva prevents clotting and the bat laps up the blood. Although men have found marks of their attacks on waking, nobody apparently has ever been awakened by the bite. Vampire bats

Below: A giant armadillo. These 4½-foot long animals shelter in burrows on the forest floor during the day, and at night use their powerful front claws to dig for food. Their varied diet includes ants, termites and other insects, spiders, worms, and snakes. They perform a useful function in destroying harmful insects and snakes.

55

NEOTROPICAL REGION
(Mexico & Central America)

© Geographical Projects

Mountain	Prairie
Coniferous forest	Mediterranean
Deciduous forest	Savanna
Temperate grassland	Tropical forest
Desert	
Semi-desert	

Brazilian tapir

Capuchin monkey
Howler monkey
Spider monkey

Capybara

Projection: Lambert's Equal Area

Scale: 1:14,300,000

Miles
0   100   200   300   400

Kilometres
0   100   200   300   400   500   600   700

120°
110°
100°
90°
30°
Rio Grande
Red
Brazos
Colorado
Pecos
Mississippi
Missis
Mississ
GUL
GUL
ME
Rio Grande
SIERRA MADRE OCCIDENTAL
SIERRA MADRE ORIENTAL
Lower California
TROPIC OF CANCER
Southern limit of transitional zone between Nearctic & Neotropical regions
Northern limit of transitional zone between Nearctic & Neotropical regions
Rio Grande de Santiago
20°
GULF OF CAMPECHE
SIERRA MADRE DEL SUR
Isthmus of Tehuantepec
GULF OF TEHUANTEPEC
P   A   C   I   F   I   C
10°
O   C   E   A   N
110°
100°

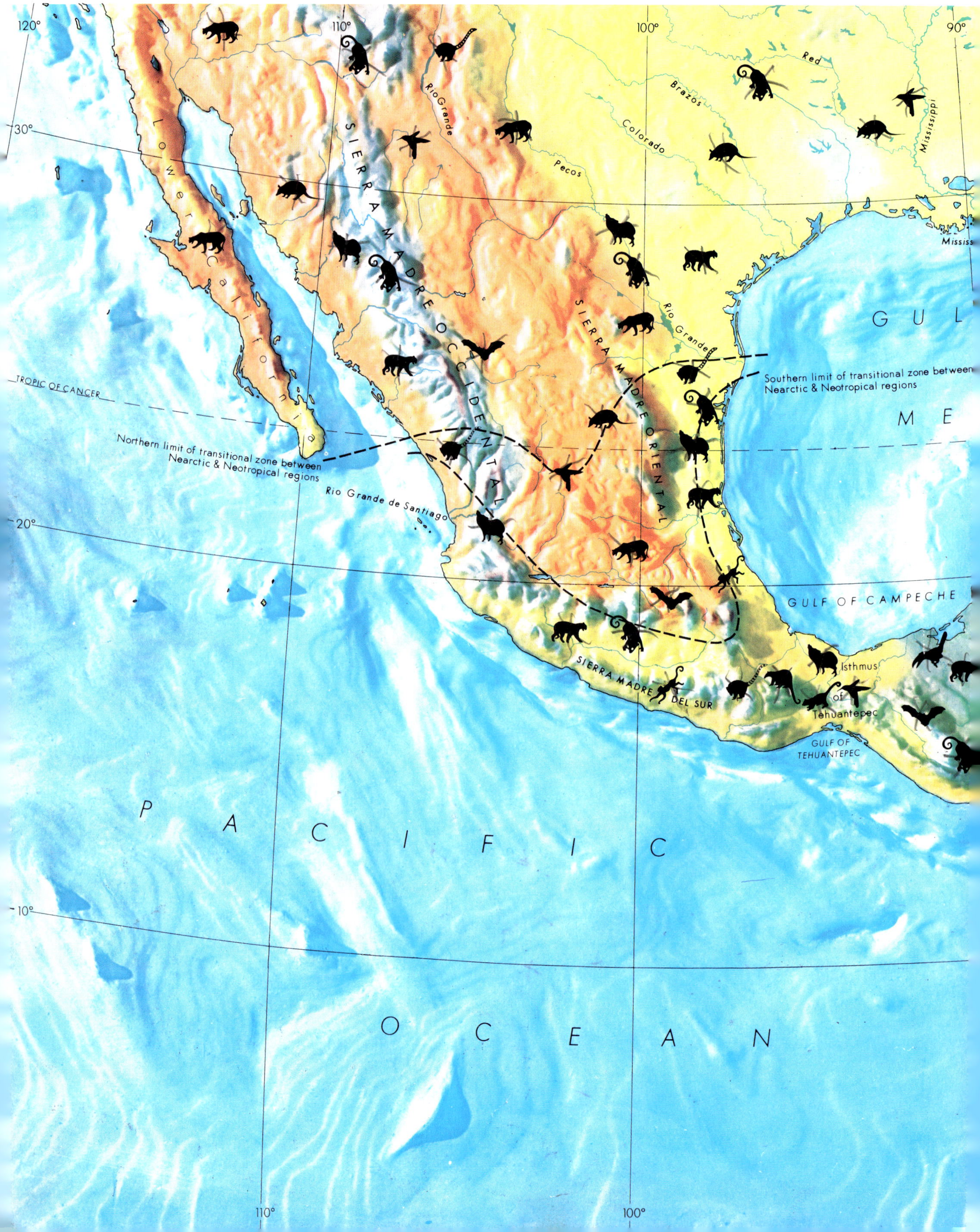

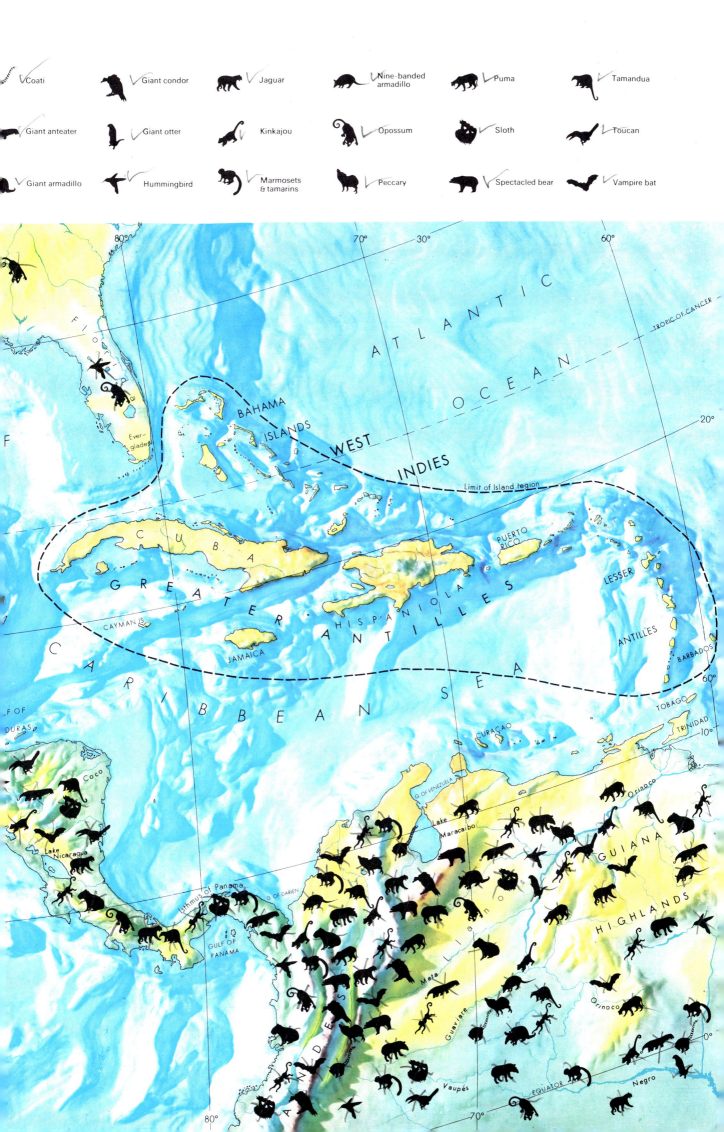

Coati
Giant condor
Jaguar
Nine-banded armadillo
Puma
Tamandua

Giant anteater
Giant otter
Kinkajou
Opossum
Sloth
Toucan

Giant armadillo
Hummingbird
Marmosets & tamarins
Peccary
Spectacled bear
Vampire bat

do not take enough blood to be dangerous to a large animal, but they do spread diseases such as rabies and horse fever.

The monkeys of South America are superficially like the monkeys of Africa and Asia because they live in a similar environment, but they are not closely related. South American monkeys generally have short snouts and widely spaced nostrils that face sideways. African and Asian monkeys, however, have downward-facing nostrils and a dog-like muzzle. Also, in most South American monkeys the tail is prehensile, whereas in African and Asian monkeys the tail (if present) is a balancing organ.

There are two families of South American monkeys. One, the Cebidae, includes the spider monkey; the howler monkey; the beautiful olive-green, white, and orange squirrel monkey; and the capuchin. The capuchin is a very intelligent animal compared with other monkeys. It can use simple tools such as stones to crack open nuts, and, like the apes, will do paintings in captivity. The monkeys of the tropical forest can be extremely noisy, particularly the howler monkeys, which defend their territories vocally like song birds. They have enlarged voice boxes that enable them to produce an amazing volume of sound.

Up in the fine twigs of the topmost branches of the trees live three unusual and shy types of cebid monkey. These are the long-haired saki, the titi, and the extraordinary, bare-headed uakari, whose face and head flushes deep red when it is angry or disturbed. These three sad-faced monkeys do not thrive well in captivity and little is known about them. The only nocturnal monkey in the world also lives in the tropical rain forest of this region. This is the douroucouli (also a cebid monkey), sometimes known as the owl or night monkey because of its large round eyes that are adapted for night vision.

The other family of monkeys in South America, called the Callithricidae, consists of the marmosets and tamarins. These are small, squirrel-sized animals and are the only monkeys equipped with claws instead of nails. Many of them have long and brightly

Left: A capybara at rest. Capybaras, at up to 4 feet long and 20 inches high, are the world's largest living rodents. They live in family groups of up to twenty individuals among dense vegetation near water. Peaceful animals with no defences against jaguars, their main predators, capybaras spend much of their time feeding on grasses and water plants. Despite their unstreamlined appearance they are excellent swimmers.

coloured fur, and this, coupled with their small size, makes them irresistible to customers in pet shops. Actually they are not easy to keep as pets, being nervous and impossible to house-train. The demand, however, is so great and the restrictions on their being hunted so few, that a plane flies between Peru and Miami each week simply to supply this trade. Eight species of monkey, including the golden lion tamarin, Goeldi's tamarin, the red uakari, and the woolly spider monkey, are at present in danger of extinction because of the pet trade.

Living on the floor of the forest and browsing on the vegetation are large herds of peccaries, distant cousins to the wild pig of Africa and Asia. The shy tapir also lives here. It is pig-like in appearance but is more closely related to the rhinoceros. The forest is also the home of three species of deer, the white-tailed deer (also found in North America), and the red and brown brockets (deer with short horns resembling young red deer stags). The many different types of hooved animals found in the forests of Africa and Asia are replaced here by unique rodents. There are guinea pigs and similar animals, such as the paca, the agouti, and the acouchi, and also the giant capybara. This four-foot-long animal spends much of its time browsing on water plants and, like the tapir and the manatee, has been largely hunted for its meat. The flesh of the capybara has a fishy taste and is sometimes sold as "salmon."

The predators that feed on these insect- and plant-eating animals include the puma and a variety of patterned cats, the largest of which is the jaguar. There is also the plain-coloured jaguarundi, which is rather otter-like in appearance with its long, sleek body and short legs. The predatory giant otter, which feeds on the smaller rodents and may be up to seven feet long, lives in the waters of the Amazon and the Orinoco rivers. It is now extremely rare, having been hunted almost to extinction for its pelt. It shares with the jaguar the doubtful honour of

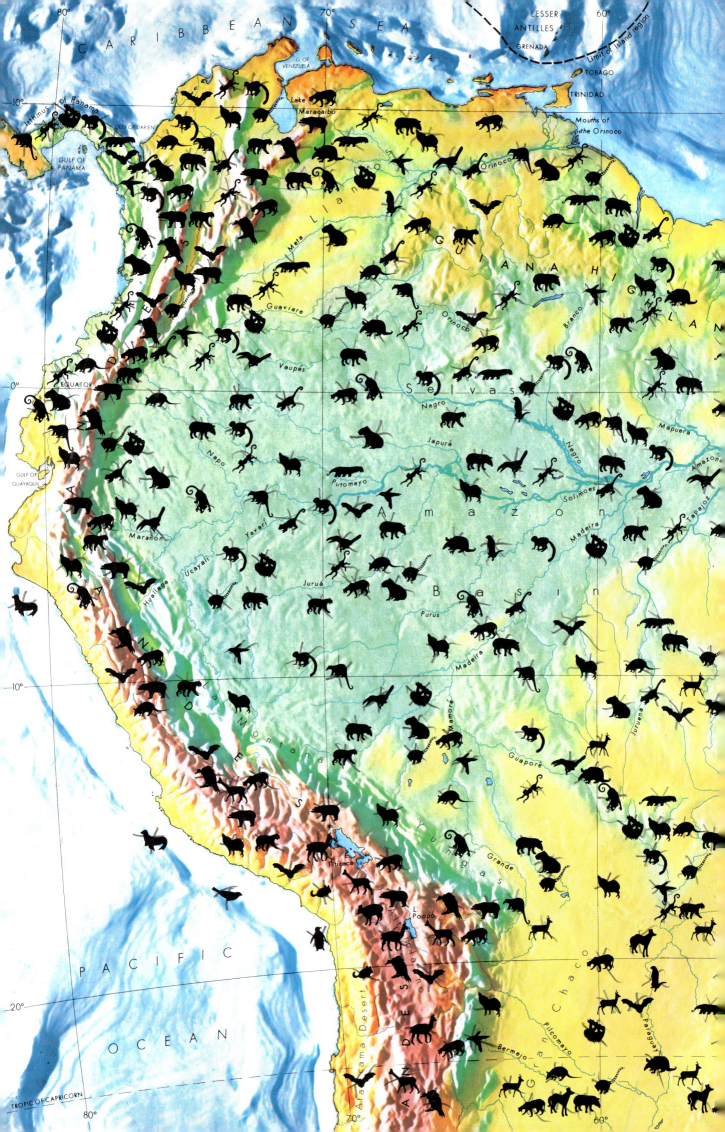

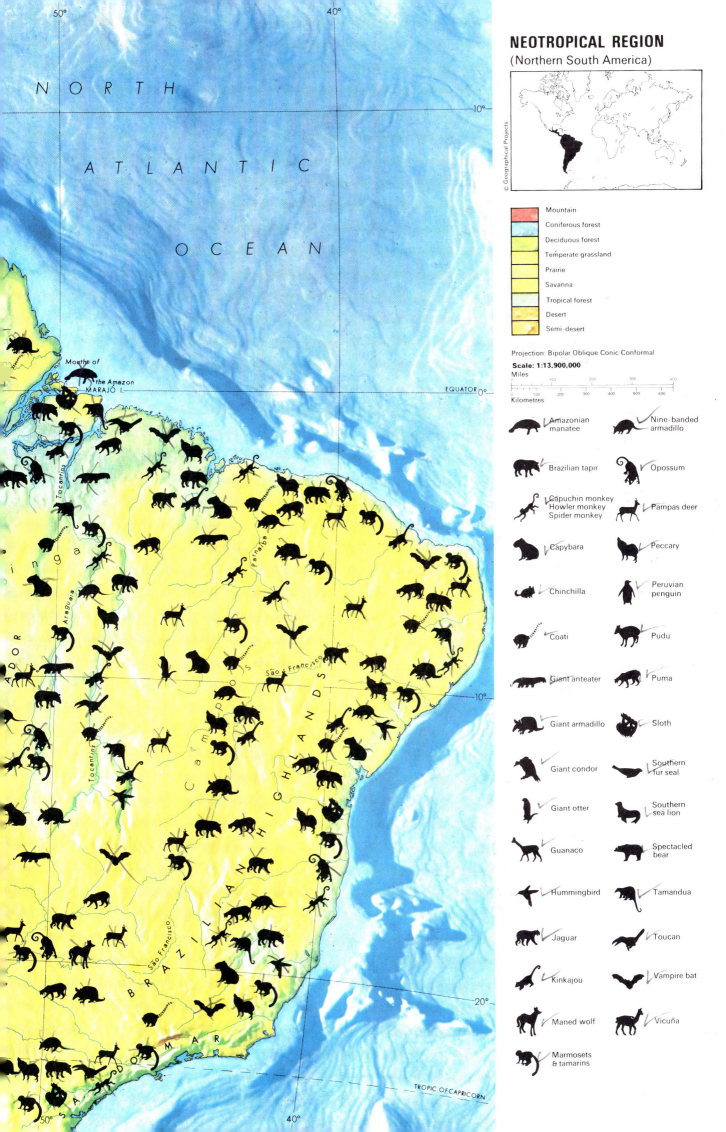

NEOTROPICAL REGION
(Northern South America)

© Geographical Projects

Mountain
Coniferous forest
Deciduous forest
Temperate grassland
Prairie
Savanna
Tropical forest
Desert
Semi-desert

Projection: Bipolar Oblique Conic Conformal

Scale: 1:13,900,000

Miles
100 200 300 400
100 200 300 400 500 600
Kilometres

Amazonian manatee
Nine-banded armadillo

Brazilian tapir
Opossum

Capuchin monkey
Howler monkey
Spider monkey
Pampas deer

Capybara
Peccary

Chinchilla
Peruvian penguin

Coati
Pudu

Giant anteater
Puma

Giant armadillo
Sloth

Giant condor
Southern fur seal

Giant otter
Southern sea lion

Guanaco
Spectacled bear

Hummingbird
Tamandua

Jaguar
Toucan

Kinkajou
Vampire bat

Maned wolf
Vicuña

Marmosets & tamarins

having the most expensive fur in the world.

Other predators include the crab-eating raccoon, which lives near the rivers, and its relatives the coati and the kinkajou, which feed on the vegetation and small animals of the forest. Kinkajous, like many of the tree-living animals of the region, have prehensile tails. Wild bush dogs, shorter in the leg than other wild dogs, hunt in packs, preying on the large rodents and small forest deer.

The best-known birds of the Neotropical region are the tiny, brightly coloured hummingbirds that can hover in front of flowers to feed on the nectar. Unlike most of the animals of this region, hummingbirds have spread northwards as far as Canada. Many other birds live in the forest including macaws, toucans, curassows, orioles, the bizarre umbrella bird with its enormous black, head-covering crest, the flame-coloured cock of the rock, and the brown hoatzin. The hoatzin is of great scientific interest because the young birds have claws on their wings, very like the earliest known fossil bird *Archaeopteryx*. If disturbed in their nest the chicks climb up the reeds or bushes using these claws and then dive into the river below. When the danger has passed they climb back to the nest.

Many snakes live in the tropical rain forest. The coral snakes, with their characteristic banded coloration, and the pit vipers are highly poisonous. The largest of the constrictors—snakes that kill by coiling around their prey—is the anaconda, which lives in swamps and rivers and may reach a length of over 25 feet. It feeds mainly on deer, peccaries, fish, and small caimans. Living with the caimans in the rivers are crocodiles and a great variety of turtles. The most hideously camouflaged of these is called the matamata, a turtle that looks more like cow dung than an animal.

There are three areas of grassy plain in the Neotropical region, the llanos of Venezuela and eastern Colombia, the campos of Brazil, and the pampas of Argentina. These vast areas of grass are now used to raise cattle, or sheep in the drier parts of Patagonia. The original animals have therefore been reduced in number or driven into other parts of the region. The Chaco,

Above: A white-lipped sicklebill, one of the 300-odd species of hummingbird, most of which live in the Neotropical region. The curved bill of this species is superbly adapted for reaching the nectaries of flowers.

Below: A jaguar stalking peccaries. Jaguars, the largest and heaviest of the American cats, live mainly in dense forest, often lurking near rivers in wait for animals that come to drink. Jaguars feed extensively on peccaries, picking off stragglers from the herds of 100 or more animals, but they also take capybaras, turtles, and fresh-water fish.

Above: A giant anteater. This solitary animal spends most of its time searching for food, walking with its nose held close to the ground. Its diet consists mainly of termites and ants, which it extracts with its long, sticky tongue, having first broken open the mounds with its powerful front claws. Giant anteaters usually run away from danger, although when cornered they will defend themselves by slashing out with their front claws. Giant anteaters are good swimmers.

an insect-infested, marshy area of north-west Argentina and western Paraguay, has become a haven for many of these plains animals.

Among the original plant-eaters of the grasslands are two deer, the marsh deer and the pampas deer, and many rodents. The rodents all belong to the cavy family, as do the capybara and other forest rodents. They include the guinea pig, the mara or Patagonian cavy, the tuco-tuco, and the viscacha. Except for the guinea pig, these are all burrowing animals. The viscacha has the habit of decorating the mouth of its burrow with collected oddments such as stones, rocks, and bones. The empty burrows are sometimes taken over by small burrowing owls.

Many of the other grassland animals are burrowers. Among these are several species of armadillo. Like the giant armadillo of the forest, these animals have small plates of bone covered with horny scales imbedded in their skin to form a covering like chain mail. Some armadillos rely on the strength of this "armour" and defend themselves by rolling up in a ball to protect their soft underparts. The three-banded armadillo does this and is known locally as the *tatu naranja*, or orange armadillo, because, rolled up, it looks like a large orange. Other armadillos take refuge by burrowing, which they can do extremely rapidly. They hold their breath as they burrow and are supposed to be able to use this ability to cross narrow rivers by walking along the bottom. The fairy armadillo, which is only five inches long and looks like a large hairy woodlouse, is highly adapted to subterranean life, having tiny eyes and ears and very large front claws.

The largest of the edentates, the six-foot-long giant anteater, is another plains animal, although it also ranges into forest areas. Like the fairy armadillo, it has powerful front claws, which it uses to tear open termite nests and to defend itself when attacked. When walking, the anteater folds these claws inwards and walks with the side and knuckles of the hand on the ground.

The birds of the plains have adapted in various ways to the absence of trees. The oven bird, for example, has overcome the problem of nesting sites by building an impregnable mound of

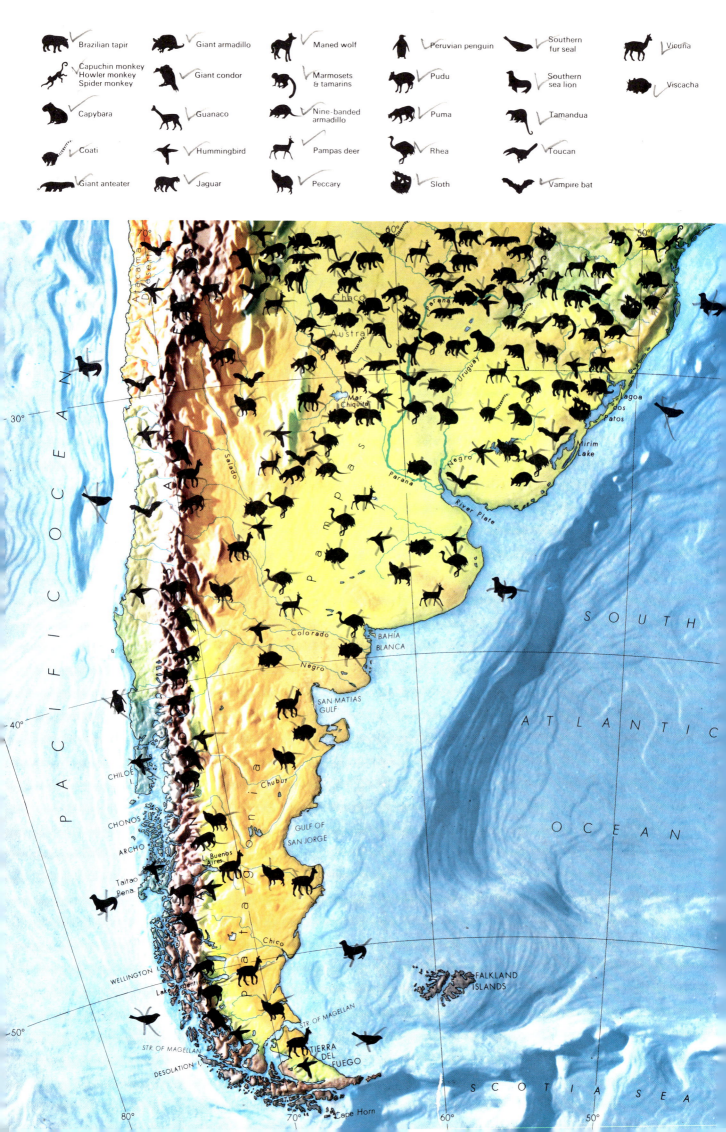

Brazilian tapir

Capuchin monkey
Howler monkey
Spider monkey

Capybara

Coati

Giant anteater

Giant armadillo

Giant condor

Guanaco

Hummingbird

Jaguar

Maned wolf

Marmosets
& tamarins

Nine-banded
armadillo

Pampas deer

Peccary

Peruvian penguin

Pudu

Puma

Rhea

Sloth

Southern
fur seal

Southern
sea lion

Tamandua

Toucan

Vampire bat

Vicuña

Viscacha

NEOTROPICAL REGION

(Southern South America)

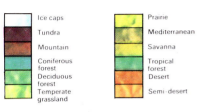

Ice caps	Prairie
Tundra	Mediterranean
Mountain	Savanna
Coniferous forest	Tropical forest
Deciduous forest	Desert
Temperate grassland	Semi-desert

Projection: Bipolar Oblique Conic Conformal

Scale: 1:13,900,000

mud on fencing posts and similar places. The bird derives its name from this construction, which is thought to look like a bread oven.

The largest bird of the grasslands is the rhea, a flightless bird of the pampas that looks very much like the ostrich, but has brownish-grey feathers. It is a very fast runner and grows to a maximum height of five and a half feet. In the breeding season the cock bird collects a flock of females and escorts each one to a communal nest (made in a depression in the ground) where up to fifty eggs may be laid. The cock then incubates the eggs.

The tinamous are also ground-living birds of the plains, although they can fly short distances if threatened. These partridge-like birds are unique to the Neotropical region and lay brightly coloured eggs with shells so hard that they look as though they are made of porcelain.

The predators of the grassy plains include pumas, coyotes, and occasionally jaguars that venture out from the forest. In the pampas grasslands live the spotted pampas cat, the pampas fox, and also the maned wolf, which looks like a fox on stilts. Maned wolves are graceful, playful creatures that feed on rodents, birds, and insects.

The Andes mountains stretch down the entire western side of South America. In some places they are extremely high and barren. On the wooded slopes of the Colombian, Ecuadorian, and Peruvian Andes lives the rare spectacled bear, the only species of bear found in the Neotropical region. Very little is known about it except that it is less carnivorous than most bears and is in danger of becoming extinct. Several deer inhabit the mountain lowlands, including the pudu, the smallest deer in the world. It stands only 15 inches high and has tiny spiky antlers. Mountain forms of guinea pigs, pacas, and viscachas, and also chinchillas, live in the Andes, although the chinchilla has been almost wiped out in the wild for its thick pelt.

Left: A common rhea. Like the ostrich, which it closely resembles, the rhea is well adapted to life on the plains. It has long, powerful legs with three toes on each foot (the ostrich has two) and can run fast, holding its neck out horizontally as it does so. Rheas feed on a wide variety of plant and animal life, including seeds, roots, leaves, insects, worms, and lizards.

65

The animals best adapted to the thin air and cold conditions of the high Andes are the guanaco and the smaller vicuña. The guanaco can live up to an altitude of 13,000 feet, and the vicuña, with its light, warm coat and large number of oxygen-carrying red blood cells, can live at 18,000 feet above sea level. The vicuña cannot be domesticated; the wild animals are rounded up, shorn, and then released. In former times, the Incas did this but reserved the fur, the warmest and lightest of all animal fur, for Inca royalty. Unfortunately the present-day smugglers of vicuña wool often take the easy, but short-sighted, step of shooting the animals and this species is now in some danger. The llama and alpaca are domesticated descendants of the guanaco. The llama is used as a pack animal and the alpaca is farmed for its wool.

The main predators of the Andes are the puma and the mountain cat. Unlike the largely nocturnal lowland puma, the mountain puma is active during the day, hunting the sheep, deer, and rodents that are its prey. Always on the lookout for a deserted kill is the Andean, or giant, condor, a huge scavenging bird with a wingspan of up to 10 feet. Other birds of the Andes include a burrowing oven bird and several species of humming-birds. Due to the lack of flowers high up in the mountains, these hummingbirds, some of which can live at an altitude of 20,000 feet, feed on insects.

On the western side of the Andes is a strip of land that often gets no rain for several consecutive years. Thus, apart from the rivers running down from the mountains, this area of Peru and northern Chile is almost barren desert. Very little is known about the animal population of the area. Among the animals recorded there are several types of lizard, a burrowing mouse, and a fox.

The promontories of rock and the islands off the western coast of South America, however, are swarming with life. The cold Humboldt, or Peruvian, Current runs northward along the coast of Chile and Peru bringing with it great quantities of plant and animal plankton and the fish that feed on it. The fish provide

Above: Vicuñas on an upland plain. Vicuñas are the smallest members of the camel family and among the most graceful of all hoofed mammals. They graze mountain grasses on semi-arid grasslands and plains at heights of between 12,000 and 18,000 feet above sea level. Each male vicuña normally leads and protects a group of five to fifteen females, warning them with an alarm trill when danger threatens.

Left: A maned wolf. This shy, solitary animal lives at the edges of swamps and in the remote areas of the plains. It is known to feed on small mammals, such as pacas and agoutis, and also on birds, reptiles, insects, and fruits. Maned wolves grow to a height of about $2\frac{1}{2}$ feet at the shoulder. In captivity they are friendly and intelligent animals.

food for fur seals and sea lions and for a spectacular number of sea birds. These include gulls, terns, fulmars, petrels, cormorants, and brown pelicans, as well as the tropical boobies and frigate birds. The birds are present in such enormous numbers that they have covered the rocks with a layer of guano so thick that 10,000 tons of it can be removed and used as fertilizer each year. The Humboldt, or Peruvian, penguin also lives on these rocky shores. The penguin is always thought of as an animal of cold regions but some species, including the Peruvian penguin and the Galápagos penguin, do live in warmer latitudes, but only where the seas are kept cold by cold currents. Other types of penguin, including the Magellanic penguin, live around Tierra del Fuego and the other islands at the southernmost tip of South America. Here the weather conditions are in sharp contrast with those of the north of the Neotropical region. Washed by the cold seas near the Antarctic Circle, these islands are a land of almost permanent cold and rain.

Animals of the Neotropical Region

The chart lists the main groups of animals found in the Neotropical region and shows if these animals occur in other regions.

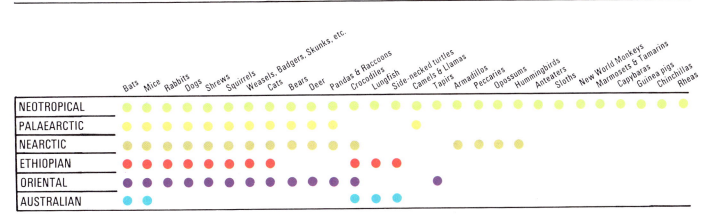

	Bats	Mice	Rabbits	Dogs	Shrews	Squirrels	Weasels, Badgers, Skunks, etc.	Cats	Bears	Deer	Pandas & Raccoons	Crocodiles	Lungfish	Side-necked turtles	Camels & Llamas	Tapirs	Armadillos	Peccaries	Opossums	Hummingbirds	Anteaters	Sloths	New World Monkeys	Marmosets & Tamarins	Capybaras	Guinea pigs	Chinchillas	Rheas
NEOTROPICAL	●	●	●	●	●	●	●	●	●	●	●	●	●	●	●	●	●	●	●	●	●	●	●	●	●	●	●	●
PALAEARCTIC	●	●	●	●	●	●	●	●	●	●				●														
NEARCTIC	●	●	●	●	●	●	●	●	●	●	●	●			●	●	●	●										
ETHIOPIAN	●	●	●	●	●	●	●	●				●	●	●														
ORIENTAL	●	●	●	●	●	●	●	●	●	●	●	●	●			●												
AUSTRALIAN	●	●										●	●	●														

5 The Ethiopian Region

Africa south of the Sahara Desert (excluding the island of Madagascar) and the south-western tip of Arabia. The Sahara Desert forms a transitional zone to the Palaearctic region.

The Ethiopian region has a great wealth of animals, having nearly as many unique families as the Neotropical region, and a great diversity within the families present, especially in the antelope family. As one would expect, the region shares a number of animals with the Palaearctic and Oriental regions. Today the Sahara Desert forms a barrier as effective as any sea, but the northern part of Africa has not always been as dry as it is now. At one time animals were able to move freely between the three regions. More surprisingly, the Ethiopian region has some reptiles, amphibians, and fish closely related to those of the Neotropical region.

Although man has plundered the wildlife of Africa since Roman times, it was not until the settling of the Cape of Good Hope by the Dutch in the 1600's that any of the species was in real danger. In the centuries that followed, particularly in the 1800's, several species, including the blue-buck and the zebra-like quagga, were made extinct. Many species are still in danger today, even though the national parks that have been set up since the late 1800's have undoubtedly saved the lives of many more.

Over vast areas of Africa, insects have dictated the pattern of man's influence on the wild animals. Where insects such as tsetse flies (which transmit sleeping sickness to man and domestic cattle) are very numerous, man cannot settle. In these areas the wild animals have survived in large numbers. In other areas, however, man has attempted to wipe out tsetse flies and sleeping sickness by clearing vegetation (some tsetse flies breed only in the shade of trees) and by indiscriminately killing off all the large wild animals that may act as a reservoir of the disease.

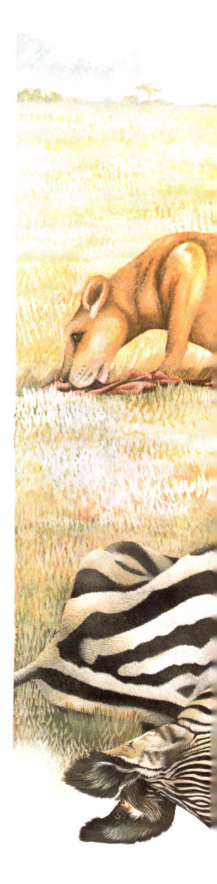

A lion with a newly killed zebra. Although the lioness probably killed the zebra, the lion is the first to feed on the carcass. The lioness is eating some of the entrails. In the background hyenas await their share of the kill.

South of the Sahara Desert are zones of semi-desert and dry grassland. This arid region stretches across Africa from coast to coast, and across Arabia, interrupted only by the Ethiopian Highlands. South of this is a belt of savanna—a zone of mixed grass and trees. The mammals that live in these dry areas of northern Africa and Arabia are, as one would expect, those that are able to survive without much water. The scimitar-horned oryx and the Arabian oryx—horse-like antelopes with long, pointed horns—are found here, as are a number of small gazelles, including the dama and the Arabian gazelles. The African wild ass, once very common here, now has a very restricted distribution, being found only in small numbers in the eastern plains of Africa.

The cheetah, the most easily tamed and most dog-like of all the cats, is the main predator of these grazing animals. It is dog-like in that it has only partially retractile claws and kills with its teeth. A cheetah chasing a gazelle will overtake it with a burst of speed, perhaps reaching 70 miles per hour for a very short period, and kill it cleanly with a single bite to the neck. Two smaller predators of the area are the golden-coloured

Right: A male ostrich. At up to 8 feet in height, ostriches are the largest of all living birds. They also lay the largest eggs of any bird, each egg measuring about 6 inches long and weighing up to 2½ pounds. Ostrich chicks develop quickly after hatching, and at one month are able to run at 35 mph. Ostriches feed mainly on fruits, seeds, and leaves, and often obtain their water supply from succulent plants.

Right: The two members of the giraffe family, a giraffe and an okapi, drawn to the same scale. Giraffes are the tallest of living mammals, growing to a height of up to 18 feet. They live in herds of as many as seventy animals on dry savannas, feeding mainly on acacia leaves. Okapis, usually solitary animals, are found only in remote parts of dense forests. They are extremely timid animals, relying on their well-developed sense of hearing to warn them of danger.

caracal lynx and the bat-eared fox. Both feed on hares, rodents, birds, and even insects.

The semi-desert areas and plains are the home of a surprising number of birds, including the largest bird in the world, the ostrich. Female ostriches are an inconspicuous brownish grey but the male is black with white wing and tail feathers. Ostriches are grazing, flightless birds and are adapted for life on the plains. Their long, strong legs and reduced number of toes—only two per foot—enable them to run at a speed of 40 miles per hour for long distances.

Many seed-eating birds, including doves, pipits, larks, and many varieties of weaver birds, live on the abundant supply of grass seeds. The weaver birds are dependent on grass in another way, for they weave a complicated spherical nest from grass to protect their eggs and young from rain and from predatory snakes.

Farther south in the savanna belt lives the hartebeest (an Afrikaans word meaning stag-like animal), a seemingly ill-proportioned creature with humped shoulders and a clumsy, rocking gait. The buffalo, another member of the cattle family, and the giant eland, a large spiral-horned antelope, are also found here. So, too, is the warthog, a large-faced pig with upcurling tusks and wart-like protuberances on its face. It has the curious habit of running with its tail sticking straight up in the air.

Three species of monkeys are found in these northern grass-lands of Africa—olive baboons, patas, and vervet monkeys. Monkeys are basically forest animals and are well suited to a climbing life, but these species have become adapted to life on the grasslands, where they feed on insects and lizards as well as on plants. The vervets, however, rarely venture far from the trees in which they spend the night. The tail, which is a balancing organ in African and Asian monkeys—not prehensile as in many South American species—has become reduced in the baboons, and in the patas monkeys it is used as a prop. These ground-living monkeys have long, powerful legs and patches of hard skin on their feet.

Using their long necks and prehensile tongues, giraffes browse the acacia trees of the savanna woodland. Once widespread throughout the savanna, giraffes are now reduced in their range, particularly in West Africa. In the arid region on the east coast known as the Horn of Africa lives another tree-browser, the gerenuk, or giraffe-necked gazelle. Like the giraffe, the gerenuk has long, thin legs and a long neck, and feeds mainly on acacia leaves. It often stands on its back legs when feeding in order to reach the higher branches. The most beautiful of the zebras, Grevy's zebra, recognizable by its narrow stripes and rounded ears, is also found in this dry, eastern area.

To the east of the main northern treeless plains and savanna belts lie the Ethiopian Highlands. Africa lacks great mountain ranges, like the Himalayas and the Andes, but has a number of isolated volcanic peaks and high tablelands. This isolation has resulted in the mountains' having a distinct collec-

Map overleaf: The wild ass symbol north of the Libyan Desert represents a small herd of the very rare Nubian wild ass. The several hundred wild asses at the Tibesti Massif, in the Sahara Desert, may be wild descendants of domestic animals.

ETHIOPIAN REGION
(North Africa)

© Geographical Projects

Mountain	Prairie	Desert
Coniferous forest	Mediterranean	Semi-desert
Deciduous forest	Savanna	Fertile lands
Temperate grassland	Tropical forest	

Projection: Lambert's Equal Area

Scale: 1:18,700,000

Miles
100   200   300   400   500   600

Kilometres
100   200   300   400   500   600   700   800   900

Aardvark

Abyssinian ibex

Addax

African buffalo

Black rhinoceros

Cheetah

Chimpanzee

Crocodile

Dugong

Elephant

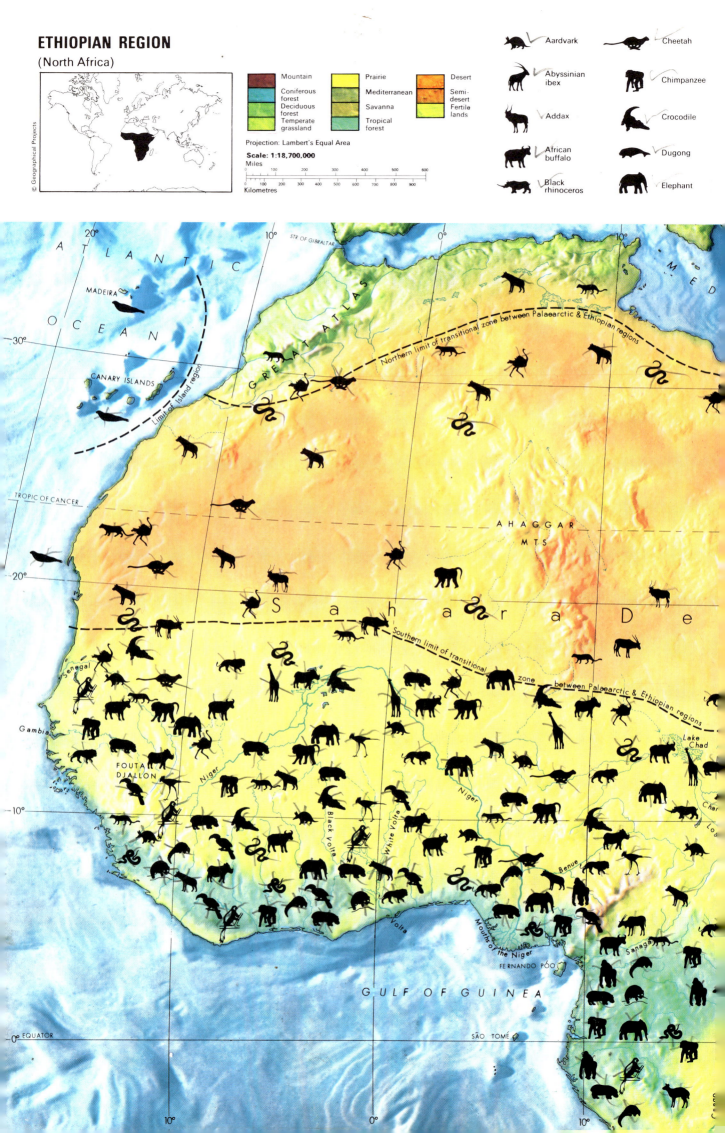

ATLANTIC OCEAN

MADEIRA

CANARY ISLANDS

Limit of Island region

GREAT ATLAS

STR. OF GIBRALTAR

MED

Northern limit of transitional zone between Palaearctic & Ethiopian regions

20°   10°   0°   10°

30°

TROPIC OF CANCER

AHAGGAR MTS

20°

Sahara De

Southern limit of transitional zone between Palaearctic & Ethiopian regions

Senegal

Gambia

FOUTA DJALLON

Niger

Black Volta

White Volta

Volta

Benue

Niger

Mouths of the Niger

FERNANDO PÓO

Lake Chad

Chari

Logone

Sanaga

10°

GULF OF GUINEA

SÃO TOMÉ

EQUATOR   0°

10°   0°   10°

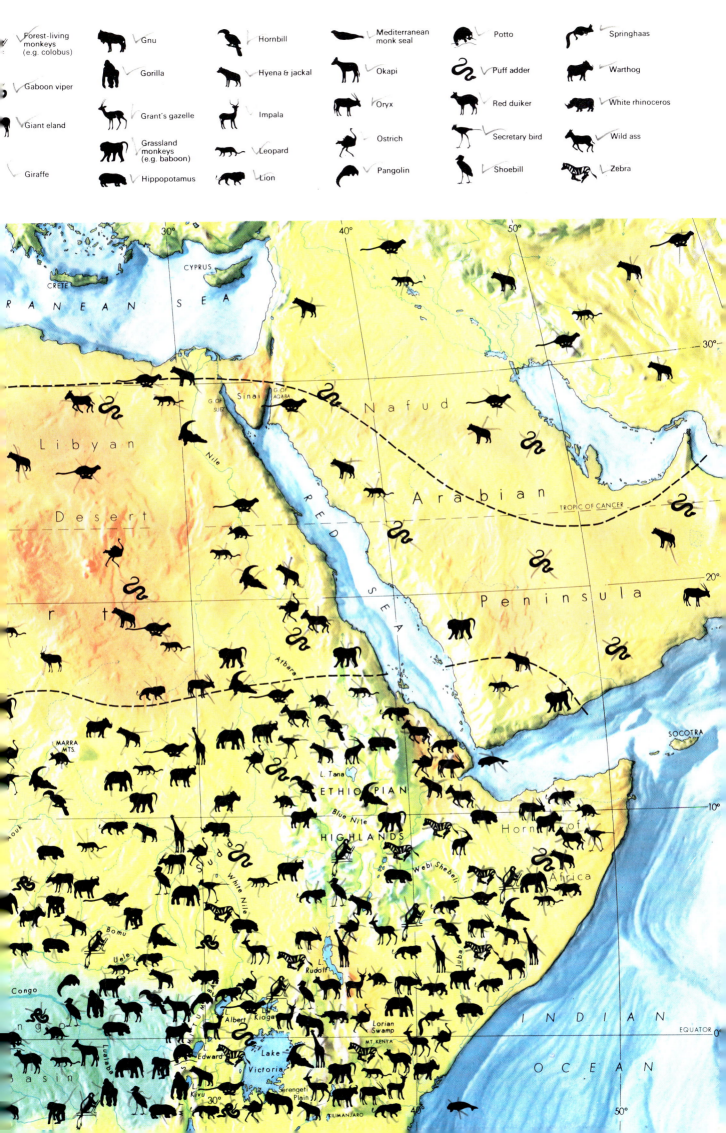

Forest-living monkeys (e.g. colobus)

Gaboon viper

Giant eland

Giraffe

Gnu

Gorilla

Grant's gazelle

Grassland monkeys (e.g. baboon)

Hippopotamus

Hornbill

Hyena & jackal

Impala

Leopard

Lion

Mediterranean monk seal

Okapi

Oryx

Ostrich

Pangolin

Potto

Puff adder

Red duiker

Secretary bird

Shoebill

Springhaas

Warthog

White rhinoceros

Wild ass

Zebra

tion of plants and animals. Heaths, lobelias, groundsels, and St. John's wort, which in temperate countries are small herbaceous plants, here grow into large bushes and trees.

The animals, too, are unique. In the Ethiopian Highlands live two ground-living monkeys, the hamadryas and gelada baboons. These are rather dog-like monkeys with long canine teeth and short tails. In both species the male (characterized by a cape of long hair on the shoulders) protects a number of females and young. Both species also have bare rump patches, and the gelada baboon has similar patches on the chest. The Ethiopian Highlands are also the home of the walia ibex, now extremely rare, and the Abyssinian wolf, otherwise known as the Simenian jackal.

The mountains farther south, which include Mount Kenya, Mount Kilimanjaro, and those mountains bordering the Rift Valley, also support an interesting animal population. The rare

Below: An adult male lowland gorilla. Gorillas are peaceful, retiring animals that attack man only if they are themselves attacked. They feed mainly on juicy plants in the forest, pulling the food apart with their hands and teeth and throwing away the unappetizing pieces. Bands of about fifteen animals, normally led by a dominant male, occupy separate areas of the forest, although the home ranges of different groups do overlap to some extent. Compared with chimpanzees, gibbons, and monkeys, gorillas are serious creatures. Only rarely have they been seen to groom each other or play.

mountain gorilla lives in the forests on the lower slopes of the mountains around Lake Kivu, west of Lake Victoria. On the hills antelope such as klipspringers and steinboks pick their way deftly over the rocks. Concealed among the rocks are colonies of brown rock hyraxes. Hyraxes, also known as dassies, are guinea pig-like animals whose nearest relative, surprisingly, is the elephant. They are the conies of the Bible.

To the south of the savanna belt lies a large area of tropical rain forest. The high canopy formed by the leaves and branches of tall, evergreen trees screens out almost all the light from the smaller trees and shrubs below, keeping the temperature and humidity at ground level very stable. Many animals thrive in this dim, moist environment. The tree-climbing monkeys of the forest include the usually brightly marked guenons (which also live in the scrublands) and the mangabeys. Both are mainly fruit-eating although they occasionally feed on insects. The leaf-eating colobus monkeys, which have been extensively hunted for their spectacularly coloured fur, live at various levels in the forest. On the forest floor live mandrills and drills. The males of these baboon-like monkeys have distinctive red and blue markings on their faces and buttocks that are used as a threat display to competing males.

The African rain forest is the home of the chimpanzee and gorilla—two of the four types of ape. (The other two are the orang-utan and the gibbon.) Apes are generally larger and more intelligent than monkeys. Their arms are long in proportion to their bodies and they have no tails. Both chimpanzees and gorillas move around in loosely organized family groups, feeding on fruit and leaves, although chimpanzees occasionally kill and eat small mammals. Both also make rough nests each night of leaves and branches and move on the next day. Adult male gorillas get so large and heavy, however, that they spend most

Above: A red colobus monkey. These leaf-eating monkeys live in the topmost branches of the forest. Below them live the black and white group of colobus monkeys and nearer the forest floor live members of the olive group. In common with monkeys of the other colobus groups, red colobus monkeys only rarely descend to the ground. They are perfectly at home in the tree-tops, crashing about noisily when intruders from other troops arrive and driving them away with loud snorts and roars. Like the spider monkeys of South America, the thumbs of colobus monkeys are poorly developed or entirely lacking.

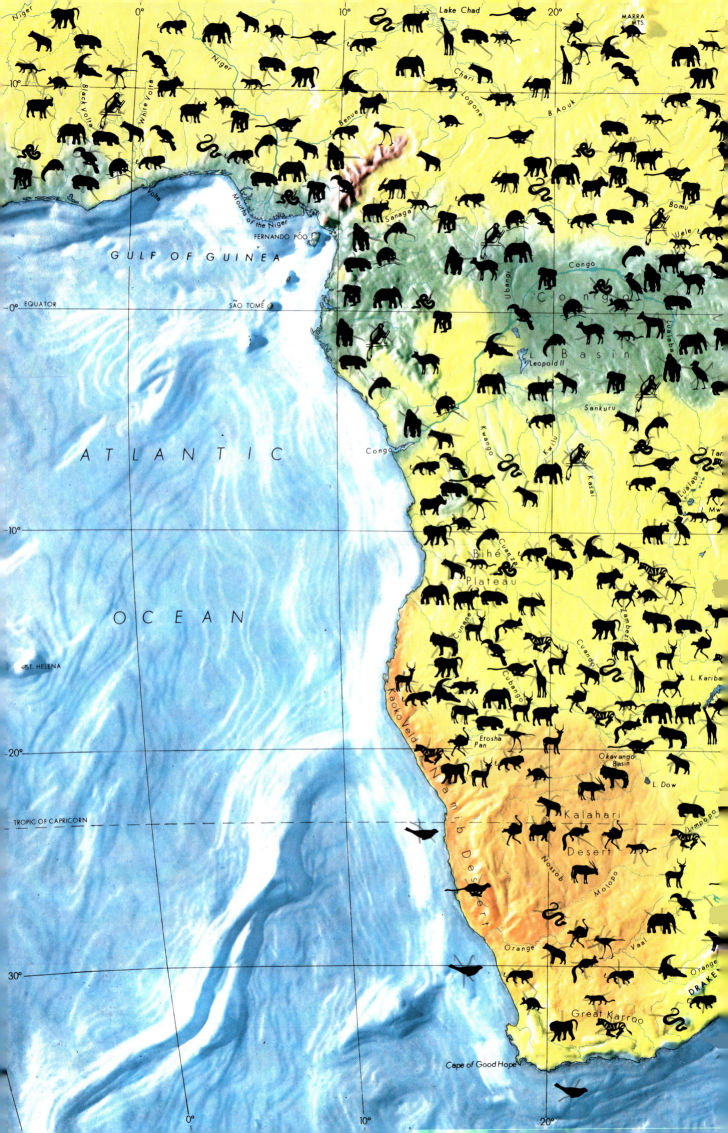

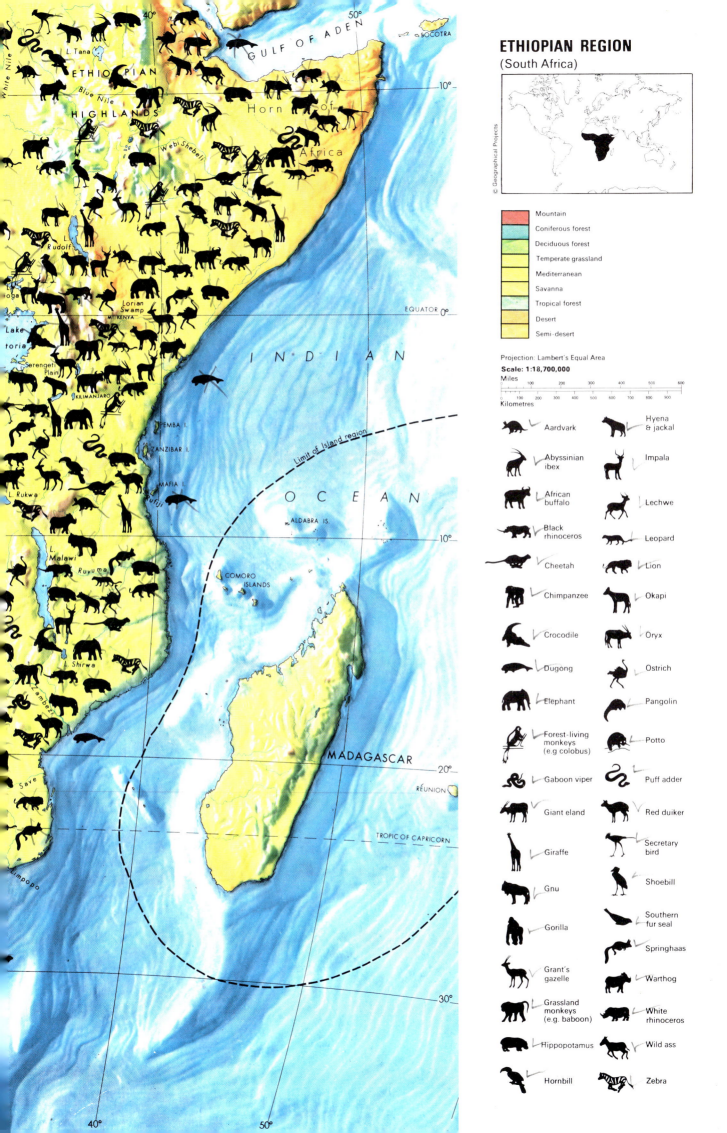

ETHIOPIAN REGION
(South Africa)

© Geographical Projects

Mountain	
Coniferous forest	
Deciduous forest	
Temperate grassland	
Mediterranean	
Savanna	
Tropical forest	
Desert	
Semi-desert	

Projection: Lambert's Equal Area

Scale: 1:18,700,000

Miles
100   200   300   400   500   600

Kilometres
100 200 300 400 500 600 700 800 900

Aardvark		Hyena & jackal	
Abyssinian ibex		Impala	
African buffalo		Lechwe	
Black rhinoceros		Leopard	
Cheetah		Lion	
Chimpanzee		Okapi	
Crocodile		Oryx	
Dugong		Ostrich	
Elephant		Pangolin	
Forest-living monkeys (e.g colobus)		Potto	
Gaboon viper		Puff adder	
Giant eland		Red duiker	
Giraffe		Secretary bird	
Gnu		Shoebill	
Gorilla		Southern fur seal	
Grant's gazelle		Springhaas	
Grassland monkeys (e.g. baboon)		Warthog	
Hippopotamus		White rhinoceros	
Hornbill		Wild ass	
		Zebra	

GULF OF ADEN   SOCOTRA

ETHIOPIAN

L. Tana

White Nile

Blue Nile

HIGHLANDS

Horn of

Webi Shebeli

Africa

L. Rudolf

oga

Lorian Swamp

MT. KENYA

EQUATOR 0°

Lake toria

Serengeti Plain

INDIAN

KILIMANJARO

PEMBA I.

ZANZIBAR I.

Limit of Island region

MAFIA I.

Rufiji

L. Rukwa

OCEAN

ALDABRA IS.

10°

L. Malawi

COMORO ISLANDS

Ruvuma

L. Shirwa

Zambezi

MADAGASCAR   20°

RÉUNION

Save

TROPIC OF CAPRICORN

Limpopo

30°

40°   50°

40°   50°

10°

10°

of their time on the ground.

Belonging to the same order of mammals as the monkeys, chimpanzees, and gorillas, but much smaller, are the bush-babies and pottos. Bush-babies, of which there are several species in Africa, some living in forest and some in savanna, make spectacular leaps from branch to branch during the night to catch the insects on which they feed. They have the curious habit of urinating on their hands and feet. This may be to help them mark out their territory and may help their sucker-like fingertips to grip. The pottos, similar in appearance to the bush-babies and with the same large eyes for night vision, also feed mainly on insects, but depend on stealth to catch their prey.

Three rather strange tree-dwellers of the African forest are the scaly-tailed flying squirrel, the scaly-tailed mouse, and the pangolin. The scaly-tailed flying squirrel uses the notch-like scales on the underside of its tail for extra grip as it clings to vertical branches. The scaly-tails are unique to Africa. Pangolins, which also occur in the Oriental region, are covered with large overlapping scales that give them the appearance of animated fir cones. If attacked they roll up like armadillos. The African tree pangolin has a prehensile tail and feeds on the termites that live in galleries in wood.

The antelopes that live on the forest floor are generally smaller than their plains-living relations, allowing them to move easily through the dense undergrowth. Most species also have patterned coats that form effective camouflage in the dappled sunlight. The bongo, for example, a forest antelope that grows to four feet high at the shoulder, has more pronounced white stripes and is smaller than its grasslands cousin, the common eland. The bushbuck, another forest antelope, is similar in size to the bongo, but the duikers are smaller, growing only to a height of about two feet. The royal, or pygmy, antelope is even smaller, being only about 12 inches high.

This tendency for the forest animals to be smaller than the equivalent grassland animals also applies to many other species. Thus the forest buffalo is relatively small, as is the pygmy hippopotamus that lives in the northern coastal regions of the forest. The okapi, an animal that was not discovered until 1901, is like a miniature edition of its savanna relative, the giraffe. Even the forest elephant is substantially smaller than the familiar bush elephant.

Other forest animals include the tiny water chevrotain, a peculiar animal that resembles the royal antelope. It is strange in that it has neither horns nor antlers and has long, spiked upper canine teeth. Chevrotains also occur in the Oriental region. There are two species of forest pig and in the litter of the forest floor live hyraxes, rats, and shrews, including the long-nosed elephant shrew, which is unique to the Ethiopian region.

The main predator of the forest is the leopard, which is a nocturnal hunter, but several smaller predators, such as the civets and genets, are also active at night. Civets and genets, graceful, cat-like creatures that belong to the same family as the mongoose, feed mainly on insects, birds, and mice. The mongoose, an animal more typical of the grasslands, is repre-

Above: Common hippopotamuses. During the day hippopotamuses rest and sleep in or near water. They feed at night, often travelling many miles on land in search of grass and other plant material, but never straying far from their river home. When diving, hippopotamuses close up their ears and slit-like nostrils.

Below: An African flap-necked chameleon. This is one of the many species of chameleons that are found throughout Africa. It grows to a total length of about 12 inches. Chameleons are well known for their ability to change colour and for the speed and accuracy with which they pick off insects with their long, sticky-tipped tongues.

sented in the forest by one species, the kusimanse.

Many of the birds of the tropical forest, like those of South America, are brightly coloured, although in the dim forest light they are less conspicuous than they appear in zoos. There are parrots, touracos (relations of the cuckoo), and hornbills. Hornbills have enormous bills and resemble the toucans of the South American forest. On the floor of the forest live the guinea fowl and the rare Congo peafowl, the only peafowl found outside the Oriental region.

The forest with its unvarying warm and humid conditions is a paradise for amphibians and reptiles. Vividly coloured tree frogs with suction-pad toes climb among the leaves, and snakes, such as the deadly mamba and the green tree viper, slide through the branches. The highly poisonous gaboon viper lives on the forest floor, its grey and brown markings making it almost invisible among the dead leaves.

The master of camouflage in the forest is the chameleon. It can change colour to melt into its background, going through shades of green, yellow, and brown. Using its pincer-like feet and prehensile tail, the chameleon can remain poised on a branch virtually still except for a slight swaying movement like the movement of leaves in the wind.

On the eastern side of the forest belt lies an area of swamps and lakes. This area includes Lake Victoria, the largest of the African lakes. The swamps, with their dense growth of papyrus reeds and Nile lettuce, are the home of two antelopes with spreading hoofs that prevent them from sinking through the vegetation. These are the sitatunga and the lechwe. Many birds live close to the water, including herons, storks, pelicans, cormorants, spoonbills, shoebills, and the sacred ibis. Nile monitors, six-foot-long lizards, feed on the eggs of these water birds and also take the eggs of crocodiles, which are common in this area.

Above: A shoebill in a papyrus swamp, waiting for fish or frogs to come within range of its large, hooked bill. These birds grow to a height of about 4 feet.

Above: A cheetah stalking a Thomson's gazelle. Cheetahs usually hunt during the day, relying on sharp eyesight rather than a sense of smell to stalk their prey.

Below: A male yellow-billed hornbill feeds his mate while a boomslang, a potential enemy to the hornbill fledglings, glides through nearby branches. The female hornbill remains in the walled-up nest until the fledglings' hunger is so great that she chips away the hardened mud and helps the male to gather food. For their own protection the fledglings rebuild the wall from the inside.

The hippopotamus also lives here, and, like the crocodile, has raised eyes and nostrils so that it can remain almost completely submerged. Unlike the crocodile, however, the hippopotamus feeds exclusively on plants.

Antelopes (such as the waterbuck), as well as buffaloes, elephants, and the white rhinoceros live around this area of swamps and lakes. The white rhinoceros is not white but grey in colour, the name being a corruption of the Afrikaans word *wyd*, meaning "wide." This refers to the white rhino's square upper lip, which is an adaptation to grazing. (The black rhinoceros is a browser and has a pointed, prehensile upper lip with which it breaks leaves off bushes.)

To the east of the lakes lies an area of grassland that merges into the thornbush that borders the Indian Ocean. This area includes the famous Serengeti National Park, a haven for game. Within the park's 5,600 square miles roam the largest herds of wild animals to be seen anywhere in the world. These herds give an impression of how the grasslands looked before the coming of the white man. Gnus (wildebeests) migrate in vast numbers during the dry season and often live together with zebras and hartebeests in mixed herds. In this way they give each other protection against predators and they do not compete for food because each type of animal eats the grass at different stages of its growth. Mixed herds of impalas, gnus, and zebras

are common and even ostriches mix with antelopes and zebras. The ever-alert ostrich, with its periscope-like neck and fine eyesight, is able to warn other animals of danger.

Graceful Thomson's and Grant's gazelles are also found in this grassland, as are black rhinos, giraffes, elephants, and lesser kudus, although these last four animals are more typical of the thornbush. The lesser kudu is a beautiful spiral-horned antelope that has been called the Apollo of antelopes.

As one would expect in an area with so many herbivorous animals there are many predators. These include cheetahs, lions, wild dogs, hyenas, jackals, and in the thornbush, leopards. The hyena, so long disliked by men for its cunning appearance and scavenging habits, has recently been found not to live entirely on animals killed by lions and cheetahs, but to hunt for itself. Lions as often finish off a hyena kill as the other way round. Vultures and marabou storks pick over anything left after these animals have taken their share.

The burrowing mole-rat, an animal unique to the Ethiopian region, performs the same weeding function on the African grasslands as the pocket gophers do on the American prairies. By its underground tunnelling and root-cropping, it also helps to keep the thornbush at bay.

Below, right: A black rhinoceros. Unlike the larger square-lipped, or white, rhinoceros, black rhinoceroses are unsociable and aggressive animals. They do not form groups, each male sharing his territory only with a female and young. Black rhinoceroses feed on twigs, buds, and leaves, which they pluck from bushes and trees with their prehensile upper lips.

The trees of the thornbush provide homes for the pretty Senegal bush-baby and a number of birds, including weaver birds and the hornbill. Hornbills have the most extraordinary nesting behaviour, for the female is walled up with the eggs inside a hole in a tree. The male bird plasters up the hole with mud so that it is too small for snakes and other predators to enter.

South of the forest region lies another vast area of savanna. Farther south is the bushveldt, an arid region of thornbush that merges into desert to the west. Like those in the north, these southern savannas have a dry season (towards the end of which fires often sweep through the tall, dry grass) and a season of rain. The numbers of grazing animals that live here, such as zebras, hartebeests, kudus, and roan and sable antelopes, have been drastically reduced by hunting and by tsetse fly control.

The large termite nests that are common in these savannas provide a source of food for the aardvark and the aardwolf. The aardwolf resembles, and is closely related to, the hyena, although it lacks the hyena's strong jaw. The aardvark (Afrikaans for "earth pig") is a curious pig-like animal with long donkey ears. Like other termite-eaters it has strong claws, a long snout, and a sticky tongue.

In the bushveldt region the Kruger National Park (over 7,000 square miles in size) still has large numbers of animals. The browsing herbivorous animals of the area include elephants, giraffes, elands, nyalas (antelopes related to elands), and greater kudus. Both black and white rhinoceroses have been reintro-duced, the original animals having been hunted to extinction, mainly for their horns. Among the grazers of this area are impalas, zebras, gnus, hartebeests, roan and sable antelopes, and in the rivers, hippopotamuses. There are no gazelles. All the

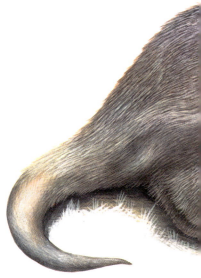

Top: A mature bull greater kudu (centre) with an immature male, and a female (without horns). Male greater kudus stand just over 4 feet high at the shoulder and are among the largest African antelopes. Resting for most of the day in scrub or in secluded spots in dark ravines, greater kudus venture out to browse and graze only in the early morning and late evening. They are extremely wary animals, relying on their acute sense of hearing to warn them of danger. They can make surprisingly high leaps, and have been seen to clear 8-foot-high bushes with ease.

typical African predators are found in this reserve. Unlike European and American parks, the African parks keep the natural predators because they weed out unhealthy herbivores and prevent overgrazing.

All the western part of southern Africa is desert or semi-desert. This region, which includes the Kalahari Desert, is very arid but does have some grass, shrubs, and even trees. Here live gnus, the common oryx (gemsbok), and the springbok, the national animal of South Africa. Springboks, graceful antelopes that once migrated across southern Africa in their millions, are now rarely seen outside game reserves. They have been hunted as a source of food and also to make way for domestic sheep, and have only just escaped extinction.

Among the smaller desert animals of this region are side-winding vipers and several types of lizard, including the web-footed gecko. The small mammals include whistling rats, ground squirrels, rock hyraxes, meerkats (a kind of mongoose), and ratels. Ratels, also known as honey badgers, are related to the skunks. They have the same warning black and white colouration and the same habit of squirting a foul-smelling liquid if attacked. Ratels are interesting for their association with a small bird, the honeyguide. This bird leads the ratel to a wild bees' nest and eats the bee larvae when the ratel has broken open the nest. Strangely, honeyguides also feed on the wax. Indeed they were first recorded in the 1500's when a priest in Mozambique noticed that they were eating his candles!

Left: An aardvark. These extremely powerful nocturnal animals live in burrows about 10 feet long and use their front digging claws to break open termite and ant nests. Aardvarks have a good sense of hearing.

Animals of the Ethiopian Region

The chart lists the main groups of animals found in the Ethiopian region and shows if these animals occur in other regions.

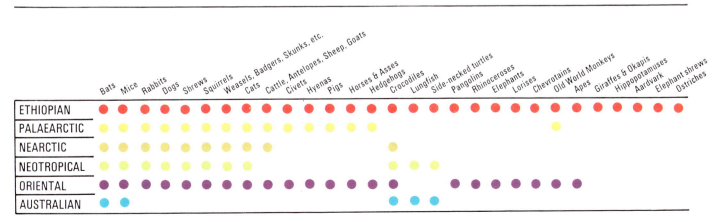

	Bats	Mice	Rabbits	Dogs	Shrews	Squirrels	Weasels, Badgers, Skunks, etc.	Cats	Cattle, Antelopes, Sheep, Goats	Civets	Hyenas	Pigs	Horses & Asses	Hedgehogs	Crocodiles	Lungfish	Side-necked turtles	Pangolins	Rhinoceroses	Elephants	Lorises	Chevrotains	Old World Monkeys	Apes	Giraffes & Okapis	Hippopotamuses	Aardvark	Elephant shrews	Ostriches
ETHIOPIAN	●	●	●	●	●	●	●	●	●	●	●	●	●	●	●	●	●	●	●	●	●	●	●	●	●	●	●	●	●
PALAEARCTIC	●	●	●	●	●	●	●	●	●	●	●	●	●	●								●							
NEARCTIC	●	●	●	●	●	●	●	●							●														
NEOTROPICAL	●	●	●	●	●	●	●	●	●							●	●												
ORIENTAL	●	●	●	●	●	●	●	●	●	●	●	●	●	●							●	●	●	●	●	●			
AUSTRALIAN	●	●													●	●	●												

6
The Oriental Region

Tropical Asia, including the islands of Ceylon and the East Indies archipelago east to Borneo and Bali, the Philippines, and some of the adjacent, smaller islands.

The Oriental region is largely an area of monsoon rains and tropical forests. Its animal population is very similar to that of tropical Africa, although it has some animals in common with the Palaearctic region, which borders its northern boundary. The spread southward from the Palaearctic region has been largely prevented by the vast ranges of the Himalayas.

Despite the cold and rarefied air there are a few specialized animals that make their homes in the Himalayas. These include the marmot, two types of wild goat—the markhor and the tahr—and the yak. Although yaks have been domesticated for many centuries by the Tibetans, who use them for transport and as a source of meat and milk, wild yaks can still be found high up in the Himalayas.

The high plateaus of Tibet are the home of the argali, a large, magnificently-horned sheep, and the Tibetan antelope, or chiru, which, like the saiga antelope, has a swollen nose. The predators that live here include the wolf, the Tibetan brown bear, and the most beautiful of all the cats, the snow leopard, with its long, dense coat spotted in shades of grey and white.

Below the high peaks of the Himalayas are dense forests of rhododendron and bamboo. At lower altitudes, in the foothills, these merge into tropical forest. This forest is the home of the Himalayan black bear, an animal that has a much wider range than its name implies, being found throughout most of the Oriental region. The takin, a large, gentle animal that resembles a gnu but is related to the musk ox of the far North, also lives high up in these forests.

In the bamboo forests to the north-east of the Himalayas, in a province of China called Szechwan, lives the giant panda. Here, too, lives the red, or lesser, panda, although this raccoon-like

A young male orang-utan. These gentle, tree-dwelling primates feed mainly on fruit and live in small family groups. Orang-utans are active during the day, clambering unhurriedly among the branches, and communicating with kissing noises. At night they sleep on platforms made with sticks and vines.

animal also ranges into the mountain forests of northern Burma and Nepal, where it feeds on a wide range of plant food. The mountain forests of Szechwan are also the home of the rare Roxellane's, or golden, monkey. This snub-nosed monkey was named after Roxellane, the beautiful wife of Suleiman the Magnificent, Sultan of Turkey in the mid-1500's.

South of the Himalayas lie the plains of the Indus and Ganges valleys, taking in East and West Pakistan and the northern provinces of India. The western plains are very dry but nevertheless support an extensive animal population.

On the Indus plains live chinkaras, or Indian gazelles, and three species of antelope, the blackbuck, the clumsy-looking nilgai, and the curious four-horned antelope. These four species are also found on the dry eastern grasslands of India. Wild boars, large sambar deer, and the smaller spotted chital deer also live on the Indus plains, as do two species of monkey, the rhesus macaque and the leaf-eating sacred langur. These animals are preyed upon by leopards and two smaller cats, the jungle cat and the African wild cat, and by wolves, foxes, and jackals. Smaller carnivores found here include the Indian mongoose and the Indian civet, a delicate creature related to the genets of Africa and to the mongoose.

Two very rare animals, the Indian wild ass (or hemione) and the Asiatic lion, live in restricted parts of this dry north-western area. Having once ranged widely in north-west India and West Pakistan, the Indian wild ass is now found mainly in an area called the Little Rann of Kutch. In historic times Asiatic lions roamed throughout the Middle East and into India, but they are now confined to the Gir forest in the Gujerat Peninsula.

The plains of the Ganges and Brahmaputra rivers in eastern India and East Pakistan are far less dry. In fact, Assam, an Indian state just to the east of the Brahmaputra River, is one of the wettest places in the world. Here the monsoon rains break with great violence and the rivers also bring vast quantities of water down from the melting Himalayan snows. In these rivers lives the gavial, or fish-eating crocodile.

The tropical forests of Assam are the main home of the Indian, or Asiatic, elephant. For many centuries the Asiatic elephant, which ranges from Ceylon to Sumatra, has been captured, tamed, and put to work in the teak forests, especially in Burma.

The one-horned Indian rhinoceros is now found only in small numbers in north-east India and Nepal, and also in preserves in Assam and Bengal. This rhino has an extremely thick skin that is formed into folds on the neck, shoulders, and hind quarters. This gives the animal an armoured appearance that is enhanced by tubercles on the skin that look rather like rivets.

The forests of these river plains also give cover to deer, the Asiatic buffalo, which is seen in its domesticated form as a draught animal all over Asia, and the gaur. Gaurs are large, cow-like animals that have a hump stretching from the shoulders to half-way down the back. They stand 6 feet 4 inches at the shoulder and may weigh up to a ton, being known locally as "bison," although they are not closely related to the American

An Asiatic, or Indian, elephant (top) and an African elephant for comparison. Asiatic elephants are smaller and lighter than the African species, reaching a maximum height of about 10 feet and a weight of 11,000 pounds. They live mainly in dense jungle, roaming in herds of up to thirty individuals, usually with an old female in the lead. Both Asiatic and African elephants feed in the evening, early morning, and at night, their diet consisting of grass, leaves, tender shoots, and fruit. Asiatic elephants are docile, intelligent animals and have long been tamed by man and put to work, particularly for moving logs in the teak forests of South East Asia.

or European bison.

South of this region of river plains India is a triangular high plateau that slopes from west to east. On the western side luxuriant monsoon forest shelters sambar and chital deer, as well as the tiny muntjac or barking deer, and the chevrotain. The sloth bear, which feeds mainly on insects, honey, and fruit, lives here, and buffalo and elephants are occasionally seen in this forest. The forest predators include wild dogs, leopards, and tigers.

The tiger is really a northern animal, coming originally from Siberia, and in hot regions must shelter from the sun. It lives in forest regions, usually near water in which it likes to lie in the heat of the day to keep cool. Apart from the occasional maneater, which is usually an animal prevented by injury from killing more agile game, the tiger shuns man. Not so the leopard, which is found not only in the forest regions but also on the savannas of eastern India. The leopard thrives near man and will often enter villages and carry off a chicken or a dog or even a sleeping man.

The forest also provides a home for many species of squirrels, including flying squirrels, and the large Malabar squirrel,

Map overleaf: The rhinoceros and tiger symbols at the western tip of Java represent small populations of the Javan races of these two animals under protection in the Udjung Kulon Reserve.

ORIENTAL REGION
(South Asia)

© Geographical Projects

Mountain
Coniferous forest
Deciduous forest
Temperate grassland

Prairie
Mediterranean
Savanna
Tropical forest

Desert
Semi-desert
Fertile lands

Projection: Lambert's Azimuthal Equal Area

Scale: 1:21,600,000

Miles
0   100   200   300   400   500   600

Kilometres
0   100 200 300 400 500 600 700 800 900 1000 1100

Asiatic lion   Elephant
Chevrotain   Flying lizard
Colugo   Gaur
Common tree-shrew
Dugong   Gavial

TROPIC OF CANCER
40°   30°   50°   60°   70°   PAMIR   80°

Plateau of Iran

HINDU KUSH

RED SEA

Arabian Peninsula

Northern limit of transitional zone between Palaearctic & Ethiopian regions

20°

PERSIAN GULF

G. OF OMAN

Sind   Thar Desert   Indus

Indie

ARABIAN SEA

G. OF KUTCH

Gir Forest

G. OF CAMBAY

Southern limit of transitional zone between Palaearctic & Ethiopian regions

GULF OF ADEN

SOCOTRA

10°

AFRICA

Deccan   Godavari

Kistna

LACCADIVE ISLANDS

0° EQUATOR

CEYLON

MALDIVE ISLANDS

Limit of Island region

10°

SEYCHELLES

I N D I A N

O

CHAGOS ARCH

MADAGASCAR

50°   60°   70°   80°

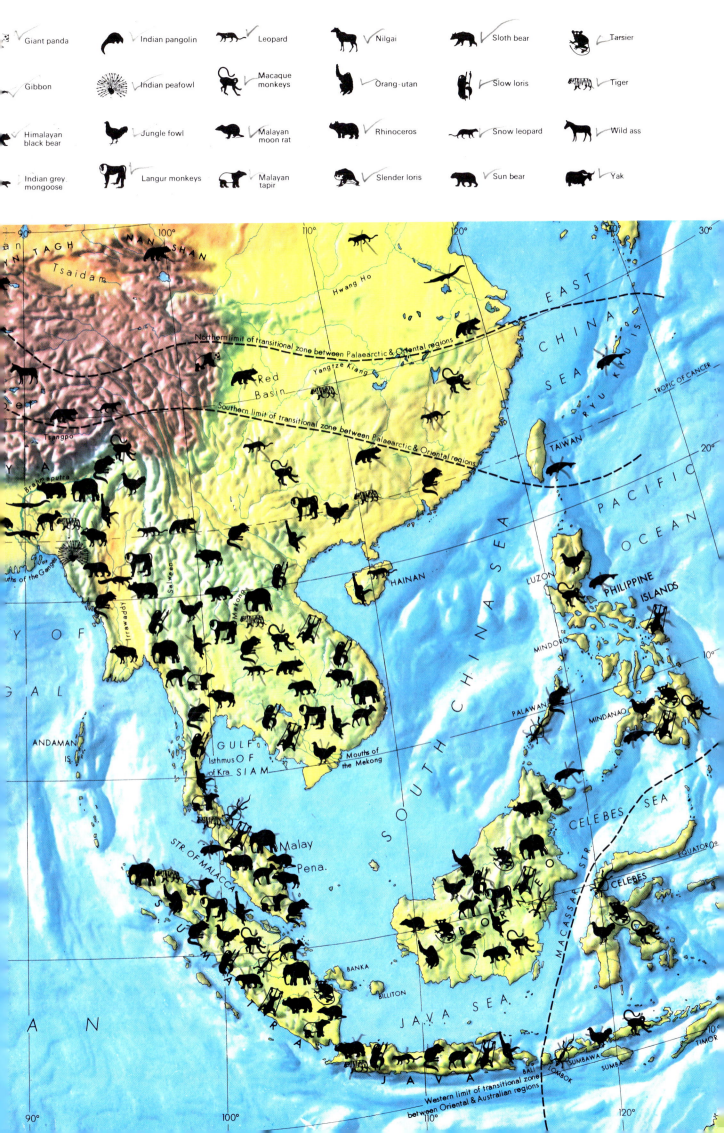

Giant panda
Gibbon
Himalayan black bear
Indian grey mongoose

Indian pangolin
Indian peafowl
Jungle fowl
Langur monkeys

Leopard
Macaque monkeys
Malayan moon rat
Malayan tapir

Nilgai
Orang-utan
Rhinoceros
Slender loris

Sloth bear
Slow loris
Snow leopard
Sun bear

Tarsier
Tiger
Wild ass
Yak

Northern limit of transitional zone between Palaearctic & Oriental regions

Southern limit of transitional zone between Palaearctic & Oriental regions

Western limit of transitional zone between Oriental & Australian regions

NAN TAGH
NAN SHAN
Tsaidam
Hwang Ho
Red Basin
Yangtze Kiang
Tsangpo
Brahmaputra
Mouths of the Ganges
Salween
Irrawaddy
Mekong
HAINAN
ANDAMAN IS.
GULF OF SIAM
Isthmus of Kra
Mouths of the Mekong
Malay Pena.
STR. OF MALACCA
SUMATRA
BANKA
BILLITON
JAVA
BALI
LOMBOK
SUMBAWA
SUMBA
TIMOR
JAVA SEA
BORNEO
CELEBES
MACASSAR STR.
CELEBES SEA
SOUTH CHINA SEA
EAST CHINA SEA
PACIFIC OCEAN
TAIWAN
RYUKYU IS.
TROPIC OF CANCER
LUZON
MINDORO
PALAWAN
MINDANAO
PHILIPPINE ISLANDS
EQUATOR
BAY OF BENGAL

which grows up to 40 inches in length. Two strange mammals of this forest are the pangolin, other species of which are also found in Africa, and the slender loris. The slender loris, like its close relation the slow loris of South East Asia, belongs to the same group of primitive primates as the bush-babies and pottos of Africa.

The birds of the western forests include doves, parakeets, mynas, and barbets. On the forest floor lives the wild peafowl together with the jungle fowl, which is the ancestor of the domestic chicken and resembles the modern bantam.

In the drier eastern lands of India the original open wood-land has been largely replaced by short-grass savanna. Preying on the gazelles and antelopes that live there are leopards, leopard cats, the common fox, wolves, hyenas, and jackals. The mongoose, which also makes its home here, feeds on small rodents and birds' eggs, as well as on its familiar prey, snakes. It is agile enough to be able to avoid the strike of even the largest cobras.

India has a number of poisonous snakes, including some pit vipers. These animals have heat-sensitive organs located in pits between the eyes and nostrils. Pit vipers use this heat-sense to hunt warm-blooded prey.

The grasslands and open woodland of Indo-China are the home of boars, deer, and many bovids (mammals of the cow family). As well as water buffalo and the huge gaur, there are banteng—which most resemble European domestic cattle and, indeed, provide excellent meat—and the rare, shy kouprey. The kouprey, the bulls of which approach the gaur in size, was, surprisingly, discovered only in 1936. Many predators are found there, these including tigers, leopards, and Asiatic wild dogs, or dholes, which, like their African cousins, hunt in packs.

Above, left: An Indian mongoose encircles a common Indian cobra. Although the cobra can attack quickly the mongoose is agile enough to avoid being bitten and usually succeeds in seizing the snake behind the head. Snakes form only a part of the mongoose's diet, which also includes insects, birds, and small mammals.

Below: A pangolin. Three species of these mammals live in the Oriental region, the largest being the Indian pangolin, which is about 3½ feet long. Pangolins locate termite and ant mounds during the night, mainly by scent.

Above: A tiger. The tiger's striped markings serve to break up its outline as it stalks its prey in long grass. Tracking its victim by sight and sound, rather than by scent, the tiger makes a sudden lunge with its fore-paws and brings the unsuspecting animal to the ground. With a bite to the throat or the back of the neck the tiger kills its prey and drags the carcass into a secluded spot before starting to feed. Tigers prey mainly on deer, pigs, gaurs, nilgais, and buffaloes, although they also eat birds, reptiles, fish, and even other tigers.

The hills and mountains of Indo-China and Malaya are covered with dense tropical forest, the leech-infested "green hell" of adventure stories. The large animals living in these forests include elephants and the Sumatran rhino. This rare, twin-horned species of rhino is smaller and has a less "armoured" appearance than the single-horned Indian rhino. Tigers and leopards also inhabit the forest, the leopard often being seen in the black (panther) form. A variety of smaller, but no less handsome, cats join their large relatives on the prowl. These cats include the clouded leopard, the golden cat, the marbled cat, and the jungle cat. These are all agile climbers, as are the civets of these forests.

The sun (or honey) bear, the smallest of all bears, is also an expert climber. It lives in forests from Burma to Borneo and feeds mainly on plants, although it also searches the tree-tops for any kind of small animals. It is particularly interested in bees' nests, which it breaks open for the honey. The trees are also the home of squirrels, flying squirrels, tropical birds, bats, pangolins, and an extraordinary animal called the colugo, or flying lemur. The colugo, which, despite its alternative name, is not a lemur, becomes active at night, gliding downhill from its resting tree to the food trees below.

The primate group is very well represented in Oriental forests. There are monkeys, apes, lorises, and the strange surprised-looking tarsiers. There are also the tree-shrews. Some naturalists do not consider these to be primates and classify them as insectivores. Indeed these pugnacious little animals do look very shrew-like, but their relatively large brains indicate that they are very like man's early ancestors, which diverged from the insectivore stock many millions of years ago.

Of the apes found in the Oriental region the gibbon is the most widespread. It occurs in Indo-China and Malaya as well as in

Above: A gavial, or gharial. These fish-eating crocodiles, which can grow up to 20 feet in length, spend most of their time submerged in water with only their eyes and nostrils above the surface. Gavials catch small fish with their small, sharp teeth by sweeping their snouts sideways through the water. They present little danger to man.

Below: A Malayan, or Asiatic, tapir. Like the tapirs found in Central and South America, Malayan tapirs are shy, docile animals that live in both dense forest and open country, usually near water. They are active during the night, emerging at dusk from their day-time hiding places to feed on water plants, and the shoots, leaves, buds, and fruits of low-growing land plants. Adult tapirs are about 7 feet long and 3 feet high at the shoulder.

the chain of islands that stretch towards Australia. The gibbon, and its close relation, the siamang, are the smallest and the least intelligent of the apes, but they are the most agile of all mammals. By swinging hand over hand they move rapidly through the forest and at the same time utter eerie, booming cries.

The Malay Peninsula south of the Isthmus of Kra and the islands that make up Malaysia, Indonesia, and the Philippines have many animals that are not found in the rest of the Oriental region. Among these is the Malayan tapir, whose nearest relation lives in South America. These two isolated groups are the remnants of a much larger distribution that once stretched across North America and Europe.

Two creatures that are unique to the southern part of the Oriental region are the otter civet and the Malayan moon rat. The otter civet resembles an otter and feeds mainly on fish, but it lacks the powerful tail and well-developed webbed feet of true

Right, top: A colugo, or flying lemur. Colugos are tree-living mammals that, despite their name, cannot fly and are not related to the lemurs. They do, however, have teeth that resemble those of lemurs and they are the best-equipped mammals for gliding flight. Colugos rest during the day, either by clinging vertically in a tree hollow or by hanging upside down from a branch. In the evening and at night they glide as much as 150 yards from tree to tree in search of the leaves, buds, and flowers on which they feed. Colugos reach a total length of about $2\frac{1}{2}$ feet.

Right, centre: A common tree-shrew. These highly active mammals, which average 15 inches from nose to tail-tip, feed during the day, mainly on insects. They are good climbers with an acute sense of smell and hearing as well as keen eyesight. Although tree-shrews have features in common with insectivores (shrews, moles, hedgehogs, etc.), they are usually classified as primates because they have certain characteristics of that group, one being their relatively large brain cases.

otters. The Malayan moon rat is a nocturnal insectivore whose narrow body and long, mobile nose allow it to search in crevices for its insect food. Strange gliding creatures, including a flying lizard, a flying frog, and a parachuting snake (the paradise tree snake), also inhabit these southern forests.

The tarsier, a small primate with enormous eyes and sucker-like fingertips, is one of the animals unique to the islands of the south-east part of the Oriental region. Tarsiers resemble the bush-babies of Africa and, like them, they can make powerful leaps among the branches.

The orang-utan (whose name means "man of the woods") at one time lived in India, China, and many of the islands, but it is now confined to Borneo and Sumatra, where it leads a family life very much like that of the chimpanzee and the gorilla. Sadly, these gentle and charming animals are declining in numbers. The island of Borneo is also the home of the proboscis monkey, the males of which develop curious, pendulous noses.

On the island of Java lives the Javan rhinoceros, a slightly smaller version of the Indian rhino. It is almost extinct, there being only about forty animals still alive. Like the Javan tiger, it is one of the most endangered large mammals of Indonesia. Farther north, the island of Mindanao in the Philippines is the only home of the rare monkey-eating eagle. In the same island group, on the island of Mindoro, lives the equally rare tamarau, a small relative of the water buffalo.

The animal and plant life of the islands at the south-east limit of the Oriental region begin to show affinities with the Australian region. It is difficult to classify these intermediate islands as they have animals typical of both regions and, in addition, some interesting, unique species. In Celebes, for example, lives the anoa, the smallest member of the cattle group that includes the water buffalo and the tamarau. The babirusa, a hairless pig with spectacular curling tusks, and several species of monkey known as Celebes monkeys, or black apes, are also unique to this island. As well as these animals, however, which are Oriental in nature, Celebes also has the cuscus and birds of paradise, which are Australasian animals. For this reason Celebes is included in a broad transitional zone that separates the Oriental and Australian regions.

Animals of the Oriental Region

The chart lists the main groups of animals found in the Oriental region and shows if these animals occur in other regions

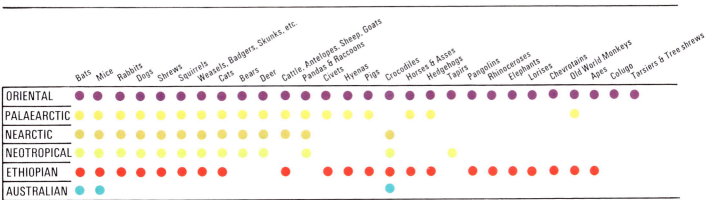

	Bats	Mice	Rabbits	Dogs	Shrews	Squirrels	Weasels, Badgers, Skunks, etc.	Cats	Bears	Deer	Cattle, Antelopes, Sheep, Goats	Pandas & Raccoons	Civets	Hyenas	Pigs	Crocodiles	Horses & Asses	Hedgehogs	Tapirs	Pangolins	Rhinoceroses	Elephants	Lorises	Chevrotains	Old World Monkeys	Apes	Colugo	Tarsiers & Tree shrews
ORIENTAL	●	●	●	●	●	●	●	●	●	●	●	●	●	●	●	●	●	●	●	●	●	●	●	●	●	●	●	●
PALAEARCTIC	●	●	●	●	●	●	●	●	●	●	●	●	●		●		●	●							●			
NEARCTIC	●	●	●	●	●	●	●	●	●	●	●	●				●		●										
NEOTROPICAL	●	●	●	●	●	●	●	●		●			●			●			●									
ETHIOPIAN	●	●	●	●	●	●	●	●			●		●	●	●	●	●	●		●	●	●	●	●	●	●		
AUSTRALIAN	●	●													●													

7 The Australian Region

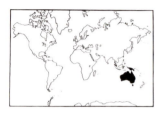

New Guinea and the neighbouring islands, Australia, and Tasmania. Celebes, the Moluccas, and the Lesser Sunda Islands form a transitional zone between the Australian and Oriental regions.

The Australian region is the only zoogeographical region that is completely cut off by sea from all other regions. It seems likely that it has been separate for something like 200 to 500 million years. This long isolation has meant that the animals of the region have evolved differently from those in the rest of the world and that Australia has many unique animals. This is particularly true of the mammals.

How do Australian mammals differ from those in other regions of the world? To answer this question we must first describe the main types of mammals. Naturalists divide mammals into three groups according to the way in which they produce their young. The largest and most successful group, and the one to which man belongs, is the placental group. In placental mammals the offspring develops inside the mother until it is more or less perfectly formed and able to cope with its environment. It is nourished inside the mother's body through an umbilical cord and a special layer, formed partly from the mother's tissue and partly from the baby's, called the *placenta* (a Latin word meaning "flat cake," which describes the shape of the tissue).

There are nearly 4000 species of placental mammals alive in the world today, of which only a few live in the Australian region.

Most of the mammals in the Australian region belong to the marsupial group (from the Latin *marsupium*, meaning "pouch" or "purse"). In marsupials the offspring is housed inside the mother's body for a very short time—about 31 days in the red kangaroo—and is not nourished through a placenta. It is born in a fairly unformed condition and crawls into the mother's pouch where it attaches itself to a teat and completes its development. The young are extremely small—a baby red kangaroo,

A female red (left) and grey kangaroo, both with young. These two species are the largest of the kangaroos, reaching a height of about 6 feet. The red kangaroo is able to live in more arid conditions than the grey, which is found mainly in the open forests of eastern Australia.

for example, weighs only half a gramme (about 1/50th of an ounce) at birth. To enable them to complete the journey from the birth canal to the pouch, the young have well-developed nervous centres and forelimbs. There are approximately 250 marsupial species and the majority of these live in the Australian region.

The third, and most primitive group, the monotremes (which means "one hole" and refers to the common urinary and genital opening), produce their young in a very strange way for mammals. They lay eggs. There are only six species of monotremes and all are unique to the Australian region. These six species are of only two types of animal: the duck-billed platypus and the echidna, or spiny anteater.

The best known of the egg-laying mammals, the duck-billed platypus, lives in mud-bottomed streams in eastern Australia and Tasmania. It feeds on fresh-water shrimps and worms and eats about half its own weight in food per day. The need to supply it with up to two pounds of worms a day is one reason why the platypus is rarely seen in captivity. When a dead platypus was first sent to England by trappers in 1798 it was greeted with scepticism. Having seen the "mermaid" made by taxidermists from half a monkey and a fish's tail, people thought that the platypus, with its duck-like bill, webbed feet, and flattened tail, was another fabricated animal.

The platypus scratches out a burrow in the bank of a stream and makes a nest of leaves. The female lays her two or three eggs and incubates them with the heat of her body. When the young hatch they are fed with milk from modified sweat glands on the mother's belly. Platypuses are fairly numerous, for they have few wild enemies. The male can protect himself with the poisonous spurs on his hind feet. Using these he can inflict a very serious and painful wound.

If the platypus looks like a cross between a duck and a beaver, then the echidna, or spiny anteater, looks part hedgehog and

Top, left: A duck-billed platypus. These egg-laying mammals, which grow to a total length of about 18 inches, live in burrows in the banks of streams and lakes. The burrow entrance is made very narrow so that the platypus squeezes its fur dry as it enters. Under water platypuses find their food mainly by the sense of touch, which is well developed in the soft skin that covers the bill.

Left: A narrow-footed, or fat-tailed, marsupial mouse. This is one of the various mouse-like marsupials that live in the Australian region and that range in total length from 3¾ to 12 inches. Like most varieties, the narrow-footed marsupial mouse, which is about 6 inches long, feeds on insects and small vertebrates, including lizards and the introduced house mouse. Marsupial mice are highly active, mainly nocturnal animals.

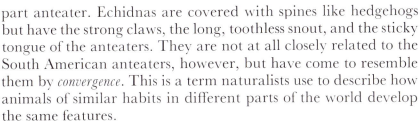

part anteater. Echidnas are covered with spines like hedgehogs but have the strong claws, the long, toothless snout, and the sticky tongue of the anteaters. They are not at all closely related to the South American anteaters, however, but have come to resemble them by *convergence*. This is a term naturalists use to describe how animals of similar habits in different parts of the world develop the same features.

Like the hedgehog, the echidna can roll itself into a ball when attacked, although it usually buries itself by digging into the earth until its soft underparts are covered. Also like the hedgehog, it finds great difficulty in keeping the skin between the spines clean and free from vermin and is continually scratching with the elongated second claw of its back feet. The female echidna, unlike the platypus, grows a marsupial-like pouch and is believed to lay the single egg straight into this pouch. Here the young echidna is fed on milk in the same way as a young platypus.

Many of the pouched animals of the Australian region have come, by convergence, to look like their placental counterparts. For example, there are marsupial "mice," "rats," and "jerboas." These are all more carnivorous than the placental animals they resemble. The marsupial anteater, the numbat, looks like a striped squirrel but has the long snout and sticky tongue necessary for its diet of termites. The most striking example of convergence is the marsupial mole, which, with its tiny eyes and well-developed front feet, is incredibly like the moles of Eurasia.

In the forest of Australia, New Guinea, and the smaller neighbouring islands live the phalangers, or possums. There are various phalangers that closely resemble American opossums, monkeys, dormice, and bush-babies. The cuscus, which lives in the tropical rain forests, is a phalanger that, both in appearance and in its slow movement, resembles the slow loris of Asia. The tiny mouse-like honey possum feeds on the nectar of flowers and has a brush tongue more typical of honey-eating birds. The flying phalangers, or gliders, have a fur-covered membrane

Top, left: A coarse-haired, or common, wombat. Like the slightly smaller hairy-nosed variety, common wombats rest during the day in or near their wide, deep burrows and emerge at night to feed on grass and roots. Wombats, which grow up to 4 feet long, dig rapidly with their front feet and use their hind feet to kick out the soil behind them. They are shy animals that often make rewarding pets.

Above: A sugar glider, or lesser gliding possum. As its name suggests, this gliding marsupial has a sweet tooth, its diet consisting mainly of buds, flowers, fruit, nectar, and sap, although it also eats insects and small birds. Sugar gliders have a well-developed pouch and grow to a total length of about 16 inches. They are active during the night, gliding as far as 60 yards between trees.

BORNEO

CELEBES

HALMAHERA

EQUATOR

JAVA SEA

BANDA SEA

BURU

SERAM

BISM

NEW GUINEA

AROE IS.

Digoel

Fly

JAVA

WETAR

TIMOR SEA

ARAFURA SEA

BALI LOMBOK SUMBAWA FLORES

SUMBA

TIMOR

TENIMBAR

Western limit of transitional zone between Oriental & Australian regions

Eastern limit of transitional zone between Oriental & Australian regions

TORRES STR.

MELVILLE I.
BATHURST I.

Arnhem Land

GROOTE EYLANDT

GULF OF CARPENTARIA

Daly

Victoria

Barkly Tableland

Mitchell

Fitzroy

Flinders

Great Sandy Desert

Ashburton

Gibson Desert

MACDONNELL RAS.

Simpson Desert

Georgina

Diamantina

TROPIC OF CAPRICORN

Gascoyne

Finke

Lake Eyre Basin

Barcoo

Bulloo

Murchison

Desert rat species solely confined to L. Eyre basin.

Warrego

Great Victoria Desert

Warburton

Lake Eyre

Coopers Cr.

Swan

Nullarbor Plain

Torrens

FLINDERS RA.

Darling

Lachlan

Short nosed species here almost extinct

GREAT AUSTRALIAN BIGHT

Gairdner

SPENCER

Murray

Murray

AUSTR. ALPS

KANGAROO I.

KING I.

BASS STR.

TASMANIA

INDIAN OCEAN

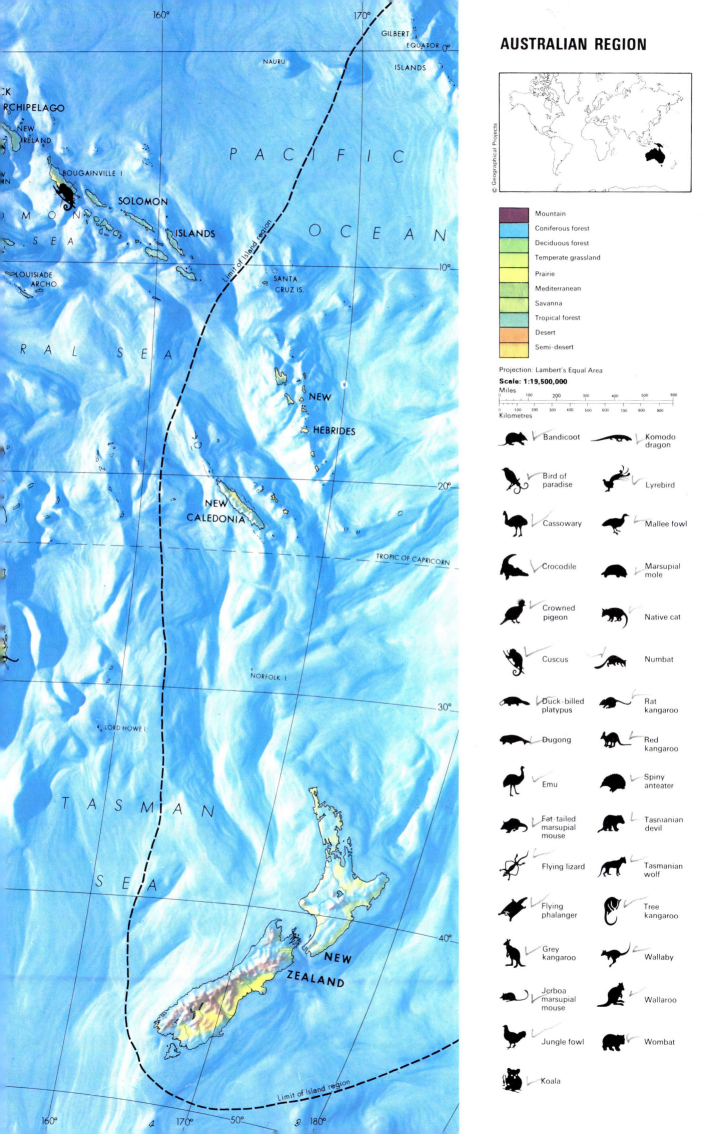

AUSTRALIAN REGION

© Geographical Projects

Mountain
Coniferous forest
Deciduous forest
Temperate grassland
Prairie
Mediterranean
Savanna
Tropical forest
Desert
Semi-desert

Projection: Lambert's Equal Area
Scale: 1:19,500,000

Miles
0   100   200   300   400   500   600

Kilometres
0 100 200 300 400 500 600 700 800 900

Bandicoot
Bird of paradise
Cassowary
Crocodile
Crowned pigeon
Cuscus
Duck-billed platypus
Dugong
Emu
Fat-tailed marsupial mouse
Flying lizard
Flying phalanger
Grey kangaroo
Jerboa marsupial mouse
Jungle fowl
Koala

Komodo dragon
Lyrebird
Mallee fowl
Marsupial mole
Native cat
Numbat
Rat kangaroo
Red kangaroo
Spiny anteater
Tasmanian devil
Tasmanian wolf
Tree kangaroo
Wallaby
Wallaroo
Wombat

between front and back legs that enables them to glide from tree to tree.

The phalangers are numerous as they have few natural enemies, being preyed upon only by domestic cats and wedge-tailed eagles. In New Zealand, where they have been introduced, these leaf-eating animals have bred so rapidly that they have had a disastrous effect on the original beech forest.

Another leaf-eating tree-dweller is the koala bear. This animal is extremely selective in its feeding habits and lives only in the eucalyptus forests of eastern Australia. Apart from making the koala smell like cough drops, its diet of eucalyptus leaves necessitates a six- to eight-foot-long appendix to aid digestion. The young develop in a backward-opening pouch and are weaned on a sort of eucalyptus leaf soup produced from the mother's anus. Just after World War I the koala was hunted almost to extinction for its fur. The public created such a fuss, however, that the fur was sold under the name of wombat fur. The koala was finally protected by law in the 1930's.

The ground-living wombat is in fact closely related to the koala. It is a burrowing, vegetarian animal very much like a large rodent, both in general appearance and also in the fact that its teeth continue to grow throughout life. Man has hunted the wombat extensively for meat and fur and it is now found only in south-east Australia. This animal also has a backward-opening pouch, obviously necessary in a burrowing animal.

The kangaroos, wallabies, and small rat kangaroos fill the

Above: A marsupial wolf, or thylacine. Now extremely rare, if not extinct, these dog-like marsupials reach a total length of about 6 feet and have a backward-opening flap of skin as a pouch. Thylacines are known to prey on kangaroos, wallabies, small mammals, and birds. They are shy, nocturnal animals, sheltering during the day in rocks or hollow logs.

Above, right: An Australian, or common, cassowary. This is one of three species that live in tropical forest in the Australian region. Like the other two types, the one-wattled and Bennett's, the Australian cassowary is adapted to running fast through dense forest. Leading with its head, which is protected by a bony helmet, the cassowary plunges into the undergrowth with the tough, uncovered quills of its wings held out to deflect entangling vines and creepers. Long, strong legs and a reduced number of toes enable the cassowary to run at 30 mph. The Australian cassowary, which grows to a maximum height of about 5 feet, feeds mainly on wild fruits and berries.

niche in nature occupied in other regions by antelope and deer. They are grazing animals, and, therefore, compete with the sheep that man has introduced. As the bush was cleared to make room for sheep farming the range and numbers of the kangaroos increased. In fact, the kangaroo increases at a greater rate than the sheep, for it can live on the hard spinifex grass that grows where the other, more appetizing, grasses have been removed by overgrazing. The red kangaroo in particular can survive in very hot, dry climates and is found throughout the central Australian desert.

The kangaroo is one of the few wild animals in the world that is still being shot in large numbers. Fifty tons of kangaroo meat are exported weekly, mostly for pet food. Some Australian naturalists think that kangaroos are in real danger of extinction, but others think that their numbers are so vast that they can withstand slaughter on this scale. Perhaps we should remember that this was also said at one time of the American bison.

In the tropical rain forests of New Guinea and northern Australia live many species of tree kangaroos. These leaf-eating animals have strong claws for climbing, a long "balancing" tail and, uncharacteristically for kangaroos, fore-limbs almost the same length as the hind limbs.

Predatory marsupials are few, possibly because of competition from the dingo dogs that were introduced by the Aborigines when they first came to the continent about 30,000 years ago. There are three main marsupial predators. The small spotted

Below: A blue-crowned pigeon. One of the several types of crowned pigeon found in the Australian region, this species measures 33 inches from bill to tail-tip and feeds mainly on seeds.

native cats or dasyures, which feed on small mammals and birds' eggs, are rather like the stoats and weasels of the Palaearctic and Nearctic regions. The other two predators are found only in Tasmania, where the dingo was never introduced. One of them, the sturdily built Tasmanian devil, was persecuted by farmers for allegedly killing chickens and is now found only in rocky and inaccessible parts of the island. It is not nearly as rare, however, as the largest predator, the dog-like marsupial wolf, or thylacine, which may in fact be already extinct. There have been no confirmed records of sightings of this animal for many years, but in 1961 and 1965 unmistakable traces of it were found in west Tasmania, so it seems likely that some are still living in remote places.

Apart from the dugong, a close relation of the manatee, and the Australian sea lion and fur seal, which live off the coast of Australia, there are only three kinds of indigenous placental mammal in the Australian region—bats, mice, and rats. However, many more have been introduced. As well as the dingo dogs already mentioned, foxes, weasels, horses, asses, pigs, camels, water buffalo, and rabbits are now all living wild in Australia.

The best-known introduction was that of the rabbit. Wild rabbits were first brought to Australia in 1859 and within three years had multiplied so greatly that they were already a pest. They had a disastrous effect on sheep farming and on the land. In some places every blade of grass was eaten and what had once been grassland became desert, for, without plant roots to bind it, the top soil was blown away. The rabbits had a doubly disastrous effect on the indigenous wild animals for not only did they compete for food, but also the foxes and weasels that were brought in to control the rabbits killed the native marsupials, which were easier prey.

The rabbit population was virtually exterminated in 1950 when 99·8% of the estimated 750 million rabbits were killed by myxomatosis, a rabbit disease deliberately introduced by man. The disease was brought from South America where the wild rabbits were immune to it. The remaining rabbits in Australia now have this immunity and the population is again increasing.

The Australian region shares many of its fish and reptiles with other regions. Lungfish, for example, which in Australia live in rivers in eastern Queensland, are also found in the Ethiopian and Neotropical regions. During the summer months, when river levels drop and stagnant pools develop, the Australian lungfish comes to the surface and breathes air. Side-necked turtles—those that draw in the head by folding the neck sideways—are also found in both the Ethiopian and Neotropical regions, as well as in the Australian region. The Australian region also has some very poisonous snakes, one type of crocodile, and several large lizards, including the mountain devils, frilled lizards, and monitor lizards. The mountain devil closely resembles another lizard of desert areas, the horned toad of America. The monitor lizards include the largest of all lizards,

Above: An adult male Raggiana, or Count Raggi's, bird of paradise. It measures 18 inches from bill to tail-tip and, like most birds of paradise, performs an elaborate display to attract the dull-coloured female bird. Birds of paradise derive their name from the once-held belief that they spend all their lives in the air, feeding on the dew of heaven. This belief arose in the early 1500's when legless skins of the birds were first brought back to Europe.

the Komodo dragon, which lives on Komodo and the neighbouring islands in the transitional zone between the Australian and Oriental regions. Komodo dragons grow to a length of nine or ten feet and will kill and eat deer.

The birds of the region include brightly coloured parrots and enormous flocks of green budgerigars that fly down to drink at waterholes. Many of the birds, such as the pigeons, also occur in the Oriental region.

Unique to the Australian region are two large, flightless birds, the emu and cassowary. The emu lives in the farmlands of western Australia and the cassowary in the forests of northern Australia, New Guinea, and the smaller islands. The emu with the kangaroo is shown in the Australian coat of arms and, paradoxically, like the kangaroo, is also being killed in large numbers. Both emus and cassowaries, like the flightless birds of South America and Africa, are fast runners with strong muscular legs and a reduced number of toes. In both cases the male incubates the eggs.

Other strange birds whose males do all the work are the brush-turkey and the mallee fowl. In both species the male bird builds a large compost heap and the female lays her eggs in this. The heat from the rotting compost incubates the eggs. This is not quite as simple and as effortless as it sounds, for in order to keep the temperature exactly at 91°F the male bird spends eleven months out of the year at the nesting site, first to prepare the mound; then, when the eggs have been laid, he keeps the temperature constant by covering and uncovering the eggs, adding leaves or hot sand to the heap, or scraping away the compost to let in cool air.

The most beautiful birds of the region are the forest-living birds of paradise, lyrebirds, and bower birds. The male birds of paradise and male lyrebirds attract mates by displaying their magnificent plumage. The male bower bird, however, achieves the same result by building a complicated bower from twigs and decorating it with brightly coloured objects.

Animals of the Australian Region

The chart lists the main groups of animals found in the Australian region and shows if these animals occur in other regions.

	Bats	Mice	Crocodiles	Lungfish	Side-necked turtles	Marsupial "mice," "cats," Tasmanian devils & wolves	Marsupial mole	Bandicoot	Phalangers, Cuscuses, Possums, Koalas	Kangaroos & Wallabies	Wombats	Spiny anteaters	Duck-billed platypuses	Cockatoos	Cassowaries	Emu	Honeysuckers	Lyrebirds	Birds of paradise
AUSTRALIAN	●	●	●	●	●	●	●	●	●	●	●	●	●	●	●	●	●	●	●
PALAEARCTIC	●	●																	
NEARCTIC	●	●	●																
NEOTROPICAL	●	●	●	●															
ETHIOPIAN	●	●	●	●	●														
ORIENTAL	●	●	●																

8 Antarctica

Antarctica is not part of any zoogeographical region. Fossil remains indicate that land animals once lived there, but severe cold now deters all except insects and marine animals.

Emperor penguins with chicks. In the background are three killer whales. Killer whales, the males of which may reach 30 feet in length, have been seen to deliberately break up ice floes so that penguins fall into the water within reach of their toothed jaws.

Antarctica is the ice-covered land mass that spans the Antarctic Circle. In contrast to the area within the Arctic Circle—the northern parts of the Palaearctic and Nearctic regions, and the frozen Arctic Ocean—Antarctica supports no land animals, except for a few mites and insects. Some of the marine animals that live around Antarctica, however, do come on to the land or the surrounding pack ice to breed.

The most familiar Antarctic animals are penguins. These are flightless birds that have taken to the sea. As they have few enemies on land and therefore little need to fly, their wings have become flippers with which they "fly" through the water. Their bodies are covered with tiny, close-set, oily feathers that keep out the water. Under these is a layer of down and a thick layer of blubber that together keep out the cold.

Many types of penguin live on the pack ice and on the sub-polar islands, but two species, the Adélie and the emperor, actually breed on the shores of the Antarctic continent. The Adélie, small black and white penguins that look as though they are wearing dress suits, live in flocks consisting of up to a million birds. This gives them a certain amount of protection against their chief enemy, the leopard seal, which lurks under the ice waiting for them to enter the water to feed. Although Adélie penguins are very efficient swimmers the leopard seals are faster.

The emperor penguins are the largest of all the species. Their large size helps them to retain heat on the extremely cold pack ice. (Large animals have a smaller surface area to bulk ratio and since heat is lost from the surface they lose heat more slowly than small animals.) Their large size, however, is also a disadvantage because the large chicks take a relatively long time to develop. In order that the chicks may be fledged in time to fish for them-

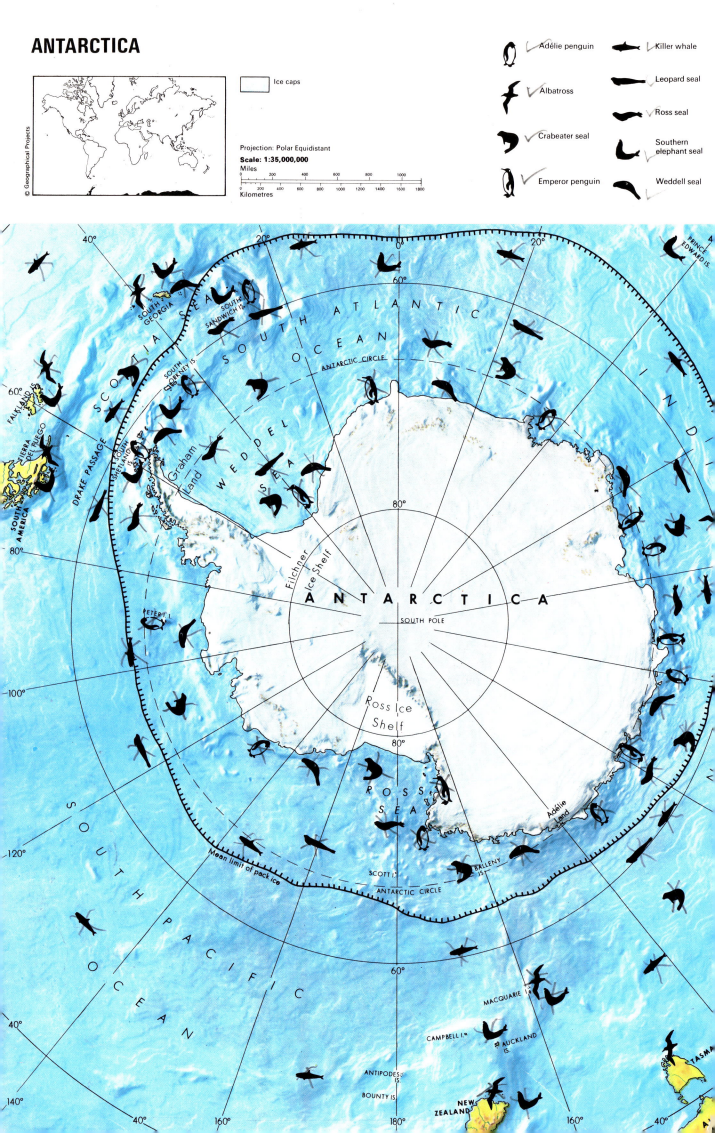

selves in the summer months, when squids and fish are most plentiful, emperor penguins lay their eggs in June and July in complete darkness and in temperatures as low as $-77.5\,^{\circ}$F. The male emperor penguin incubates the single egg against a bare patch on his belly, balancing it on his feet and covering it with a flap of skin and feathers.

The parents take it in turn to brood the chick, one brooding it while the mate travels to the open sea to feed and fill its crop with food for the chick. Both parents have such a strong brooding instinct that they will brood a dead chick or even an egg-shaped lump of ice! If a chick is put down on the ice for an instant it may be torn to pieces as several penguins fight to brood it.

Penguin chicks are also in danger from skuas, large sea birds that not only kill young penguins but also drive adult penguins off their eggs, which they eat. Petrels and fulmars, sea birds related to the albatross, also invade penguin colonies—but to scavenge rather than to kill. They feed on dead penguin chicks and dead seal pups, although, like skuas, they also take fish and crustaceans in the open sea or from breaks in the pack ice.

The wandering albatross, the largest of all sea birds, with a wingspan of up to 11 feet, glides thousands of miles across the south polar seas and nests on the islands just outside the Antarctic Circle.

As well as the leopard seal there are three other true Antarctic seals, the crabeater seal, the Ross seal, and the Weddell seal. The enormous southern elephant seal, with its characteristic trunk-like nose, visits Antarctic waters but breeds on islands north of the Antarctic Circle. These seals all live in a narrow circle of sea around Antarctica. They are able to live at such close quarters to each other because their habits and food differ.

The leopard seal lives in the water around the edge of the pack ice and feeds on carrion and fish, as well as on penguins. The crabeater seal lives among the pack ice, feeding on krill (small shrimp-like animals present in incredibly large numbers). The rare Ross seal, however, swims under the pack ice and with its large eyes searches out fish and squid in the semi-darkness. The Weddell seal sticks to coastal waters, coming up to breathe at blowholes in the ice. This seal, which feeds on the fish and squid of deep waters, can dive to a depth of 1,800 feet, staying under water for up to an hour at a time. During these dives, the Weddell seal, in common with other seals, conserves oxygen by cutting off blood to all but the vital organs.

Many whales, including the huge blue whale, the humpback, and the fin whale, spend the summer months around Antarctica feeding on the dense concentration of krill, which they sieve out of the water with the vast whalebone combs in their mouths. These three whalebone whales are among those most prized by whalers for their abundant supply of oil-yielding blubber.

One of the most common whales of Antarctica is the killer whale. This belongs to the toothed whales, a group that includes the sperm whales, dolphins, and porpoises. Killer whales are voracious animals, feeding not only on seals and penguins but also on other whales. They hunt by sight, leaping out of the water to see what prey is on nearby ice. Like other whales, they also use an echo-location system while under water.

Below: A Weddell seal. These coastal seals grow to a length of about 10 feet, and spend much of the time in the water, especially during winter. They dive deep in search of fish, squids, and crustaceans, and can remain submerged for up to an hour. Cracks in the ice are used as breathing holes and a Weddell seal returning to the surface will often clear an iced-over breathing hole by sawing through the ice with its strong canine teeth.

9
Islands

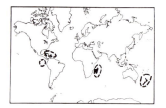

The islands considered in this chapter—New Zealand, Madagascar, the West Indies, and the Galápagos group—have animal populations that reveal the unique nature of island life.

A takahé. This flightless bird of New Zealand was once thought to be extinct, but, in 1948, it was rediscovered in South Island and is now protected by law. It makes its nest between tussocks of grass and lays not more than four eggs a year—some of which may be infertile. The takahé, of which between 200 and 300 are known to exist, is the badge bird of the New Zealand Ornithological Society.

When the English naturalist Alfred Russel Wallace divided the world into zoogeographical regions in the 1870's he encountered a problem. Certain islands had animal populations that were either so restricted or so bizarre that he found it difficult to classify them according to his major world regions. Old, continental islands such as New Zealand and Madagascar were difficult to place. These islands were probably connected with a large land mass at one point in time but have been separate for so long that the island animals have evolved differently from those on the parent continent. Oceanic islands such as the Greater Antilles (part of the West Indies) and the Galápagos Islands in the Pacific Ocean also present problems for the zoo-geographer. These islands were formed from undersea volcanoes and have never had a connection with any continent. The animals that live on them have crossed the sea from the nearest land masses.

The islands of New Zealand have been isolated from other land masses for at least 200 million years. It seems certain that they separated, if indeed they were ever joined to another land mass, before the age of mammals, for New Zealand has no native mammals, except for two types of bat, which could have reached it by flying.

This lack of mammals meant that ground habitats that would normally be occupied by cattle, deer, pigs, and similar animals were open to birds. Thus, in the absence of competition, many kinds of flightless birds have evolved in New Zealand. The best known of these is the kiwi, New Zealand's national animal. The kiwi's wings are so small that they are completely lost in the brown, hair-like feathers on its body. Kiwis are nocturnal,

probing the ground with their long beaks and using their keen sense of smell (unusual in birds) to find insects and worms.

The brightly coloured takahé, another of New Zealand's flightless birds, was once thought to have become extinct, but, in 1948, it was rediscovered in a remote wooded valley on South Island. The takahé is now rigorously protected. The almost flightless owl-parrot, or kakapo, has only just escaped extinction, having been hunted for food.

These surviving species pale into insignificance when compared to New Zealand's original population of flightless birds. When the Polynesians first settled on the islands in about A.D. 950 there were approximately twenty-two species of giant flightless birds living in the forests and grasslands. These were ostrich-like birds, some standing up to 12 feet in height. They undoubtedly made good eating and those that were not exterminated by the Polynesians fell prey to the Maoris who arrived between 1150 and 1350. However, it seems that at least one species of moa (as the Maoris called these giant birds) survived until the mid-1700's. There was also a flightless wren, the only flightless perching bird known to exist, but with the introduction of predators such as stoats it was wiped out along with many other native animals.

Of the many flying birds found in New Zealand, some are unique to the islands but most are also found in Australia, or have a wide distribution. The kea, a large green parrot that is found only in New Zealand, has developed new eating habits since the introduction of sheep to the islands. Originally, the kea fed mainly on leaves, fruit, and seeds. Now it is reported to attack live sheep, tearing away the skin of the rump in order to reach the kidney fat.

With its temperate climate and its forests with brooks and streams, New Zealand is well suited to amphibians. Paradoxically, only one amphibian is found on the islands. This is a frog (*Leiopelma*) whose nearest relative lives in the ponds and rivers of North America. This animal is a survivor of a group that had a wide distribution about 150 million years ago but that has now largely become extinct.

This also applies to the tuatara, one of the reptiles native to New Zealand. The tuatara is a large, beaked lizard-like reptile found only on a few islets off the New Zealand coast. In common with many lizards, the tuatara has a third, or pineal, eye in the centre of its head. This is better developed in the adult tuatara than in any other animal, but it is not known precisely what function it serves.

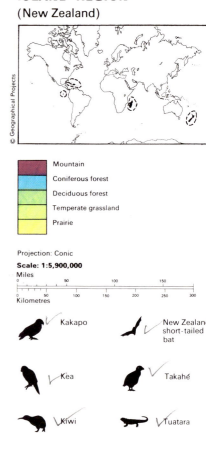

ISLAND REGION
(New Zealand)

Mountain
Coniferous forest
Deciduous forest
Temperate grassland
Prairie

Projection: Conic
Scale: 1:5,900,000
Miles
Kilometres

Kakapo

New Zealand short-tailed bat

Kea

Takahé

Kiwi

Tuatara

New Zealand frog

Map opposite: The takahé is restricted to wild country in the Murchison Mountains west of Lake Te Anau and also in the Kepler Mountains north of Lake Manapouri. This area of south-west South Island is part of the Fiordland National Park.

Right: A tuatara. Although outwardly resembling modern lizards, the tuatara is in fact the only surviving member of an ancient group of reptiles that became extinct at least 100 million years ago. Tuataras, which grow very slowly to a maximum length of about 2 feet, live in burrows, either of their own making or those made by petrels and other birds. They feed mainly on insects, although they occasionally eat petrel eggs and chicks.

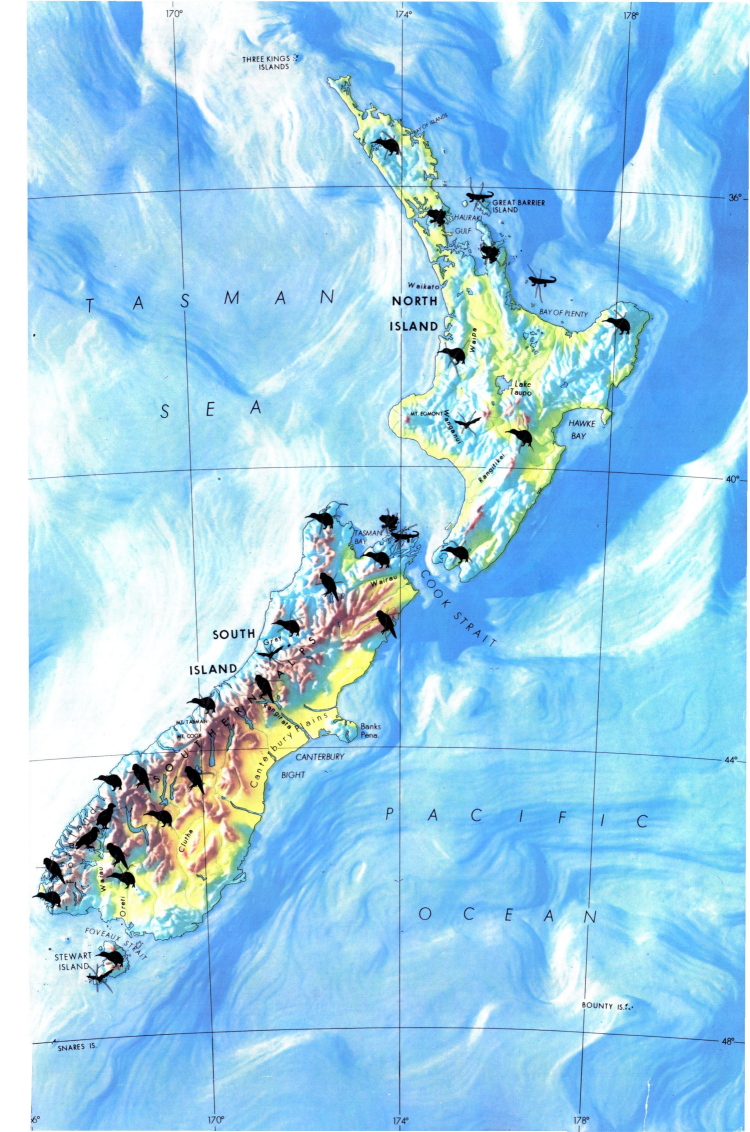

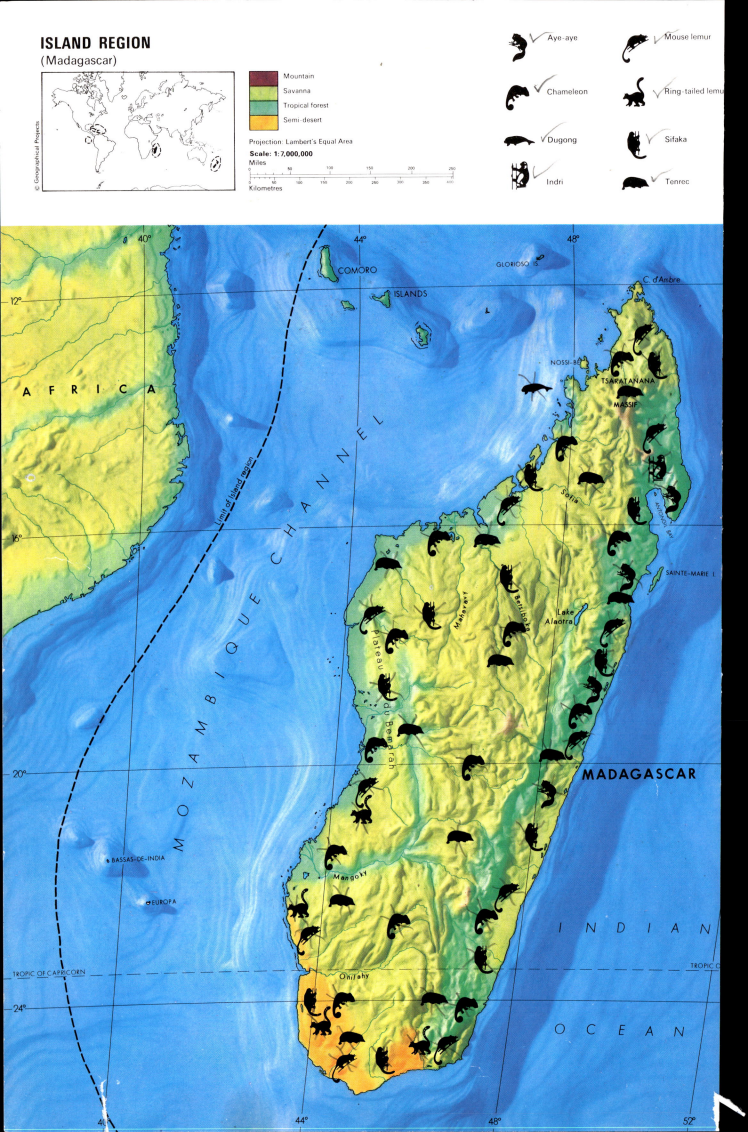

ISLAND REGION
(Madagascar)

© Geographical Projects

Mountain
Savanna
Tropical forest
Semi-desert

Projection: Lambert's Equal Area

Scale: 1:7,000,000

Miles
0   50   100   150   200   250

Kilometres
0   50   100   150   200   250   300   350   400

Aye-aye
Chameleon
Dugong
Indri

Mouse lemur
Ring-tailed lemur
Sifaka
Tenrec

COMORO

ISLANDS

GLORIOSO IS.

C. d'Ambre

AFRICA

NOSSI-BÉ

TSARATANANA

MASSIF

ANTONGIL BAY

SAINTE-MARIE I.

Sofia

Limit of island region

M O Z A M B I Q U E   C H A N N E L

Mahavavy

Betsiboka

Lake Alaotra

Plateau du Bemarah

MADAGASCAR

BASSAS-DE-INDIA

EUROPA

Mangoky

I N D I A N

TROPIC OF CAPRICORN

TROPIC O

Onilahy

O C E A N

Above: An aye-aye. Unlike true lemurs the aye-aye has claws on all digits except the first toe. The aye-aye, whose name comes from the native word describing its occasional cry, grows to a nose-to-tail-tip length of about 3 feet. Below: An indri. Measuring about 2½ feet from head to rump, indris are the largest of the lemurs. They feed on leaves, fruit, and flowers, and utter weird, almost human-sounding, calls.

Although Madagascar lies close to the eastern coast of Africa, the great difference between its animals and those of the Ethiopian region show that it has been an island for a very long time. Unlike New Zealand, Madagascar is the home of many mammals, most of them unique to the island. These include a family of insectivores (insect-eating mammals), the tenrecs. They are found only in Madagascar but have grown to resemble insectivores and rodents of other lands. There are shrew- and mole-like tenrecs, a beaver-like water tenrec with webbed feet and a flattened tail, and a spiny tenrec that looks exactly like a miniature hedgehog.

The most beautiful animals of Madagascar are the lemurs. They are generally the size of monkeys, with fox-like faces and long tails. Many live in the forest, although the ring-tailed lemur lives among the rocks. Like their relations, the bush-babies of Africa and the lorises of Asia, they have nails on their fingers and toes but retain one claw on the second toe of each foot, which they use for scratching.

There are also tiny mouse-like lemurs and two large lemurs, the black-and-white indri and the sifaka. Both these animals are rare and are now protected. They have probably been saved from extinction by the awe and superstition that they have engendered in the inhabitants of Madagascar. Myths and legends describe the indri, for example, as a sun-worshipper. This has no doubt arisen from the indri's habit of sunbathing with its arms held up towards the sun so that its body gets maximum exposure. The indri is also said to catch spears thrown at it and hurl them back with unfailing accuracy.

An even stranger primate of the forests of Madagascar is the nocturnal aye-aye, a squirrel-like animal with coarse dark brown hair and large eyes. It is now extremely rare and lives only in the north-eastern coastal forests of the island. The aye-aye feeds mainly on wood-boring insects and their larvae. It uses its well-developed senses of smell and hearing to locate the larvae and then it tears the rotten wood away with its rodent-like front teeth and digs them out of the tunnels with the long, thin third finger of its highly specialized hands.

Although the reptile population of Madagascar generally resembles that of Africa (the most common reptile of the island is the chameleon, for example), there are some significant differences. Madagascar has no poisonous snakes nor any of the agamid type of lizards that are common in Africa and Asia. Instead the island has the iguanid type of lizards that are found in America.

East of Central America lie the West Indies, a large group of islands most of which have probably never been connected to the mainland. Considering how close the islands are to the American continent, it is amazing how much the native animals of the West Indies differ from those on the mainland.

The most significant islands for the naturalist are the four biggest: Cuba, Hispaniola (Haiti and the Dominican Republic), Jamaica, and Puerto Rico, which together are known as the Greater Antilles. Here can be found most of the few land mammals native to the West Indies. On Cuba and Hispaniola live the two species (one to each island) of a curious and now rare insectivorous mammal, the solenodon. Superficially it resembles the tenrecs of Madagascar and the moon rats of Asia and, like them, it is nocturnal, hiding in hollow logs or limestone caves during the day. After dark the solenodon forages for insects,

ISLAND REGION
(West Indies)

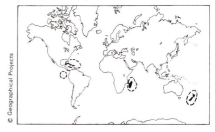

© Geographical Projects

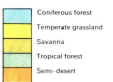

Coniferous forest

Temperate grassland

Savanna

Tropical forest

Semi-desert

Projection: Lambert's Conformal Conic

Scale: 1:10,150,000

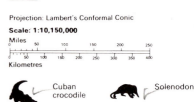

Miles

Kilometres

Cuban crocodile

Solenodon

Hutia

West Indian monk seal

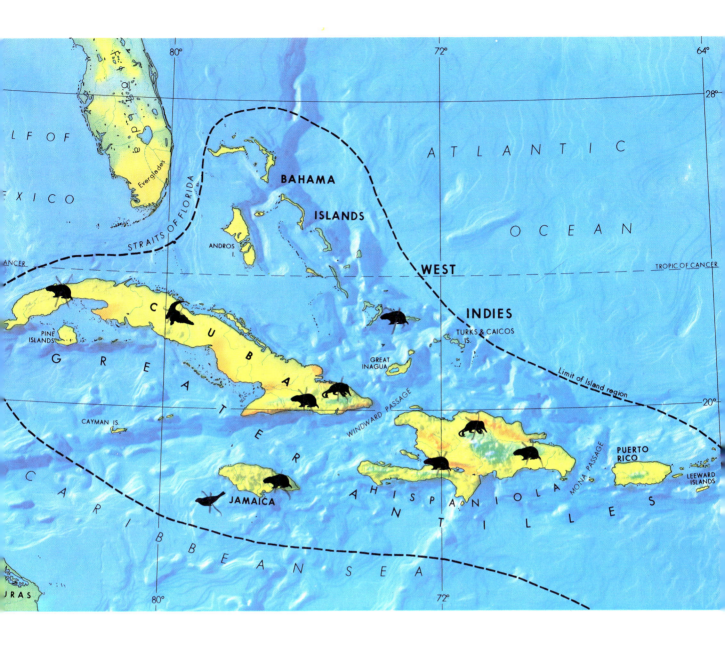

Map above: The West Indian monk seal symbol represents the small number of these extremely rare animals that may still survive in Jamaican waters and on remote islets off the coast of Yucatan. Left: A Hispaniolan solenodon. Like the longer-haired species found on Cuba, the Hispaniolan solenodon feeds at night, mainly on insects and other invertebrates, reptiles, and fruit. Solenodons, which grow to a total length of about 20 inches, walk on their clawed toes with a waddling gait that makes them follow a zigzag course. The Hispaniolan species is known to produce poisonous saliva, the ducts of the salivary glands leading to grooves in the second lower incisor teeth. Solenodons are now rare and in danger of becoming extinct, mainly because they have a low reproductive rate and are preyed upon by mongooses and other introduced animals.

using its sensitive snout and sharp hearing to locate its food.

The other principal native mammals, apart from the numerous bats, are rodents. Unique to the islands is the hutia, a rodent that is related to the large, rat-like nutria, or coypu, found in central and southern South America. Hutias are now rare and on some islands are in danger of extinction. This is mainly because they are preyed upon by the mongooses introduced to kill the European rats and mice that have over-run the islands.

The amphibians (very few), reptiles, and birds of the West Indies represent a mixture of North, Central, and South American forms. The American crocodile, for example, occurs on Hispaniola, Jamaica, and also on Cuba, where it competes with the smaller, and now rare, Cuban crocodile. Strangely, some Central and South American families of birds, such as the tinamous and toucans, have not crossed the comparatively small water gap and have not colonized the islands.

Six hundred miles west of South America lie the Galápagos Islands, a group of fifteen volcanic, oceanic islands that are world famous for their association with Charles Darwin and his theory of evolution. The animals of the Galápagos Islands are derived mostly from those that have swum, flown, or drifted across from America. The animal life is not particularly rich, there are only two mammals—a bat and a mouse, for example—but it is interesting because the animals that have settled on the islands have adapted in various ways to their new-found environment.

This is particularly true of Darwin's finches, a group made up of different species that have developed from one type of finch that reached the islands, probably from South America. When Darwin visited the Galápagos Islands in the mid-1830's he found that each island had its own race of finches with beaks fitted to feed on the most abundant food of the island. This sparked off in Darwin's mind the idea that animals gradually evolve into different forms as they become adapted, over many generations, to new living conditions. Later Darwin used his observations of the Galápagos finches as part of the evidence for the theories he set out in his famous book *On the Origin of Species*.

The Galápagos Islands are also famous for their unique reptiles. The dragon-like marine iguanas that reached the islands from South America have given rise to land iguanas, which are now rare. The marine iguanas are still common on the island shores, where they feed on seaweed. Like the land iguanas, they may reach a length of over three feet.

When the Spaniards discovered the Galápagos Islands in the 1500's they found many giant land tortoises and named the islands after these animals (*galápago* is the old Spanish name for tortoise). These tortoises (which are also found on the island of Tonga in the Pacific and Aldabra in the Indian Ocean) are among the longest-lived of all creatures, reaching an age of over one hundred years. Several of the races of tortoises found on the

ISLAND REGION
(Galápagos Islands)

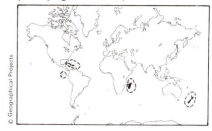

© Geographical Projects

Savanna
Semi-desert

Projection: Mercator
Scale: 1:950,000
Miles

Kilometres

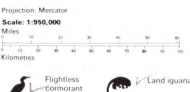
Flightless cormorant

Land iguana

Galápagos penguin

Marine iguana

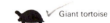
Giant tortoise

Below: A Galápagos land iguana and several giant tortoises. Male land iguanas defend a territory against other males, fights often breaking out if ritual displays do not deter intruders. The giant tortoises, the largest of which may weigh 500 pounds, vary slightly in the shape of their shells from island to island. These slow-moving reptiles spend a lot of the time half submerged in water, but are able to survive drought because they store water. Both the land iguanas and the tortoises of the Galápagos Islands have been reduced in number, and on some islands wiped out, by man and his domestic animals.

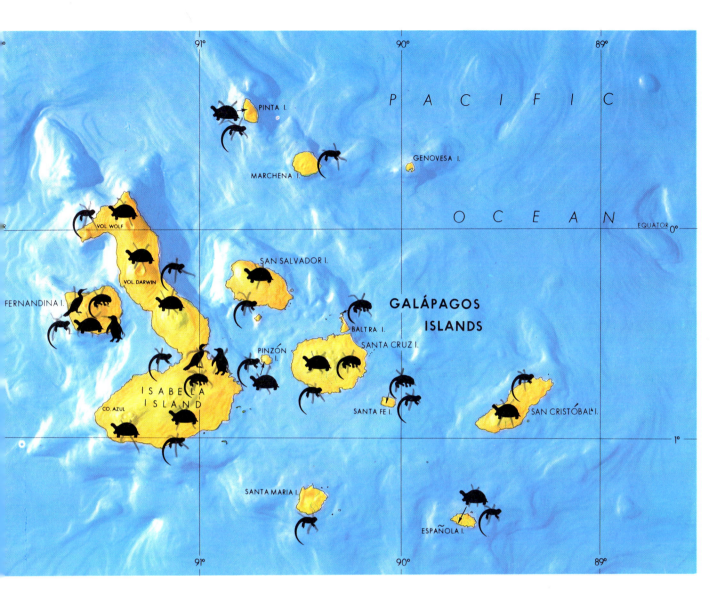

The following labels appear on the map:

PACIFIC OCEAN

PINTA I.

MARCHENA I.

GENOVESA I.

EQUATOR 0°

VOL WOLF

SAN SALVADOR I.

VOL DARWIN

FERNANDINA I.

GALÁPAGOS ISLANDS

BALTRA I.

SANTA CRUZ I.

PINZÓN I.

ISABELA ISLAND

CO. AZUL

SANTA FE I.

SAN CRISTÓBAL I.

SANTA MARIA I.

ESPAÑOLA I.

91° 90° 89°

Map above: The giant tortoise symbols on the largest island (Isabela) represent five races, each centred around one of the five volcanoes, which were originally separate islands. Although the volcanoes are now joined by land, the tortoise populations have not inter-mingled because almost impenetrable lava plain lies between them.

Galápagos Islands are now extinct, however, mainly because of hunting by man and the destruction of the vegetation by introduced animals. The surviving tortoises are now rare and protected by law. The ancestors of the Galápagos tortoises are thought to have reached the islands by floating across from America.

It is difficult to imagine that animals, such as the Galápagos tortoises, can colonize remote islands, but a recent example proves that it is possible. In 1883 the volcanic island of Krakatau (midway between Sumatra and Java) erupted and all plant and animal life on it was destroyed. In 1908, twenty-five years after the eruption, a team of naturalists made a brief survey of the island and found it grass-covered and supporting a varied animal population. This included one species of lizard, a gecko, snails, spiders, woodlice and many other insects, as well as thirteen species of bird. Another survey was made in 1921 and by then there was a forest on the island and three times as many animal species as in 1908. These included a rat, two kinds of bat, a python, and an earthworm. Twelve years later, in 1933, the survey team found a crocodile on the island and a total of over 1,000 species of animals living there. In the face of this evidence the idea of giant tortoises bobbing over vast stretches of the ocean to their present island home seems more plausible.

Wildlife Parks of the World

✱ Selected by M. J. Ross-Macdonald,
Editor of *The World Wildlife Guide*
(Threshold, London ; Viking, New York)

Not shown on map
Antarctica (open to
tourism since 1968)

BSa	=	Bird Sanctuary	NR	= Nature Reserve
CA	=	Conservation Area	NWR	= National Wildlife Refuge
FR	=	Fauna Reserve	P	= Park
GR	=	Game Reserve	PA	= Protection Area
GSa	=	Game Sanctuary	PP	= Provincial Park
MA	=	Management Area	R	= Refuge
NC	=	Nature Conservation Territory	Res	= Reserve
NM	=	National Monument	Sa	= Sanctuary
NNR	=	National Nature Reserve	SGR	= State Game Reserve
NP	=	National Park	SP	= State Park

✱ A representative selection from among the world's many thousand parks, reserves, and sanctuaries. All the parks shown here play an important part in conservation and offer the visitor opportunities to observe the wildlife they contain. Political boundaries are drawn only where they are necessary to show the location of parks.

New Zealand
1 Fiordland NP
2 Mt Cook NP
3 Westland NP
4 Arthur's Pass NP
5 Tasman NP
6 Egmont NP
7 Tongariro NP
8 Urewera NP
Fiji
9 Naqaranibuluti NC
10 Ravilevu NR

UNITED STATES OF AMERCIA
Pacific States
11 Hawaii Volcanoes NP (Hi)
12 Haleakala NP (Hi)
13 Aleutian Is NWR (Alaska)
14 Clarence Rhode NWR (Alaska)
15 Mt McKinley NP (Alaska)
16 Kenai National Moose Range (Alaska)
17 Olympic NP (Wash)
18 Mt Rainier NP (Wash)
19 Crater Lake NP (Ore)
20 Klamath NWR (Ore)
21 Sacramento NWR (Calif)
22 Yosemite NP (Calif)
23 Kern-Pixley NWR (Calif)
Mountain States
24 Desert NWR (Nev)
25 Bryce Canyon NP (Utah)
26 Bear River NWR (Utah)
27 National Bison Range (Mont)
28 Glacier NP (Mont)
29 Bowdoin NWR (Mont)
30 Medicine Lake NWR (Mont)
31 Yellowstone NP (Wyo)
32 Grand Teton NP and National
   Elk Refuge (Wyo)
33 Monte Vista NWR (Colo)
The Southwest
34 Saguaro NM (Ariz)
35 Grand Canyon NP (Ariz)
36 Wichita Mts NWR (Okla)
37 Tishomingo NWR (Okla)
38 Aransas NWR (Tex)
39 Laguna Atascosa NWR (Tex)
40 Santa Anna NWR (Tex)
41 Big Bend NP (Tex)
42 Bosque Apache NWR (N Mex)
The Midwest
43 Quivira NWR (Kan)
44 Kirwin NWR (Kan)
45 Fort Niobrara NWR (Nebr)
46 Wind Cave NP (S Dak)
47 Sand Lake NWR (S Dak)
48 Lake Ilo NWR (N Dak)
49 Slade NWR (N Dak)
50 Des Lacs NWR (N Dak)
51 J Clark Salyer NWR (N Dak)
52 Isle Royale NP (Mich)
53 Kirtland's Warbler Management
   Area (Mich)
54 Ottawa NWR (Ohio)
55 Spring Mill SP (Ind)
56 Horicon NWR (Wis)
57 Upper Mississippi NWR (Minn)
58 Mark Twain NWR (Ill)
59 Flint Hills NWR (Kan)
The South
60 Holla Bend NWR (Ark)
61 Reelfoot NWR (Tenn)
62 Gt Smoky Mts NP (Tenn)
63 Shenandoah NP (Va)
64 Chincoteague NWR (Va)
65 Hungry Mother SP (Va)
66 Wheeler NWR (Ala)
67 Eufaula NWR (Ala)
68 Okefenokee NWR (Ga/Fla)
69 Merritt Is NWR (Fla)
70 Loxahatchee NWR (Fla)
71 Everglades NP (Fla)
72 JN "Ding" Darling NWR (Fla)
73 Delta NWR (La)
74 Yazoo NWR (Miss)
75 Sabine NWR (La)
The Northeast
76 Erie NWR (Pa)
77 Iroquois NWR (NY)
78 Missisquoi NWR (Vt)
79 Moose Horn NWR (Me)
80 Morton NWR (NY)
81 Blackwater NWR (Md)

Mexico
96 Cumbres de Monterrey NP
97 El Cogorron NP
98 Nevado de Toluca NP
99 Nevado de Toluca NP
100 Ixtacihuatl-Popocatepetl NP
Guatemala
101 Tikal NP
102 El Pino NP

Ecuador
122 Galápagos Islands

Peru
120 Callao guano stacks
121 Nazca Vicuña Reserve

Canada
82 Glacier NP
83 Kootenay NP
84 Banff NP
85 Jasper NP
86 Cypress Hills PP
87 Prince Albert NP
88 Wood Buffalo NP
89 Duck Hills PP
90 Quetico PP
91 Algonquin PP
92 Laurentides PP
93 Gaspesian PP
94 Fundy NP
95 Cape Breton Highlands NP

Venezuela
103 Sierra Nevada de Merida NP
104 Henri Pittier NP
105 El Avila NP
106 Guatopo NP
107 Canaima NP
Guyana
108 Kaieteur Falls NP

Brazil
109 Paulo Afonso NP
110 Sooretama NP
111 Rio Dole Sa
112 Iguassu Falls NP (see also Argentina)
Argentina
113 Finca El Rey NP
     Iguacu Falls NP (also
        ... Paraguay)
     ... ... NP
116 Nahue ...api NP
     ... A...ces NP
     ... Moreno NP
115 ... eciares NP

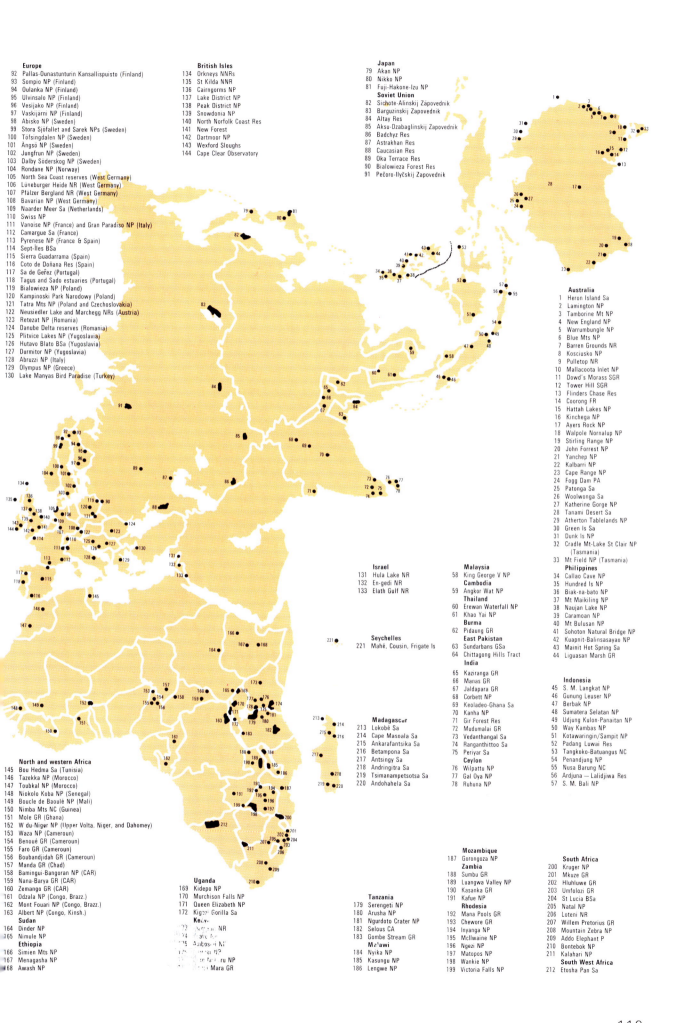

Europe

92 Pallas-Ounasturturin Kansallispuisto (Finland)
93 Sompio NP (Finland)
94 Oulanka NP (Finland)
95 Ulvinsalo NP (Finland)
96 Vesijako NP (Finland)
97 Vaskijärni NP (Finland)
98 Abisko NP (Sweden)
99 Stora Sjöfallet and Sarek NPs (Sweden)
100 Töfsingdalen NP (Sweden)
101 Angsö NP (Sweden)
102 Jungfrun NP (Sweden)
103 Dalby Söderskog NP (Sweden)
104 Rondane NP (Norway)
105 North Sea Coast reserves (West Germany)
106 Lüneburger Heide NR (West Germany)
107 Pfälzer Bergland NR (West Germany)
108 Bavarian NP (West Germany)
109 Naarder Meer Sa (Netherlands)
110 Swiss NP
111 Vanoise NP (France) and Gran Paradiso NP (Italy)
112 Camargue Sa (France)
113 Pyrenese NP (France & Spain)
114 Sept-Iles BSa
115 Sierra Guadarrama (Spain)
116 Coto de Doñana Res (Spain)
117 Sa de Gerêz (Portugal)
118 Tagus and Sado estuaries (Portugal)
119 Bialowieza NP (Poland)
120 Kampinoski Park Narodowy (Poland)
121 Tatra Mts NP (Poland and Czechoslovakia)
122 Neusiedler Lake and Marchegg NRs (Austria)
123 Retezat NP (Romania)
124 Danube Delta reserves (Romania)
125 Plitvice Lakes NP (Yugoslavia)
126 Hutavo Blato BSa (Yugoslavia)
127 Durmitor NP (Yugoslavia)
128 Abruzzi NP (Italy)
129 Olympus NP (Greece)
130 Lake Manyas Bird Paradise (Turkey)

British Isles

134 Orkneys NNRs
135 St Kilda NNR
136 Cairngorms NP
137 Lake District NP
138 Peak District NP
139 Snowdonia NP
140 North Norfolk Coast Res
141 New Forest
142 Dartmoor NP
143 Wexford Sloughs
144 Cape Clear Observatory

Japan

79 Akan NP
80 Nikko NP
81 Fuji-Hakone-Izu NP

Soviet Union

82 Sichote-Alinskij Zapovednik
83 Barguzinskij Zapovednik
84 Altay Res
85 Aksu-Dzabaglinskij Zapovednik
86 Badchyz Res
87 Astrakhan Res
88 Caucasian Res
89 Oka Terrace Res
90 Bialowieza Forest Res
91 Pečoro-Ilyčskij Zapovednik

Australia

1 Heron Island Sa
2 Lamington NP
3 Tamborine Mt NP
4 New England NP
5 Warrumbungle NP
6 Blue Mts NP
7 Barren Grounds NR
8 Kosciusko NP
9 Pulletop NR
10 Mallacoota Inlet NP
11 Dowd's Morass SGR
12 Tower Hill SGR
13 Flinders Chase Res
14 Coorong FR
15 Hattah Lakes NP
16 Kinchega NP
17 Ayers Rock NP
18 Walpole Nornalup NP
19 Stirling Range NP
20 John Forrest NP
21 Yanchep NP
22 Kalbarri NP
23 Cape Range NP
24 Fogg Dam PA
25 Patonga Sa
26 Woolwonga Sa
27 Katherine Gorge NP
28 Tanami Desert Sa
29 Atherton Tablelands NP
30 Green Is Sa
31 Dunk Is NP
32 Cradle Mt-Lake St Clair NP
   (Tasmania)
33 Mt Field NP (Tasmania)

Philippines

34 Callao Cave NP
35 Hundred Is NP
36 Biak-na-bato NP
37 Mt Maikiling NP
38 Naujan Lake NP
39 Caramoan NP
40 Mt Bulusan NP
41 Sohoton Natural Bridge NP
42 Kuapnit-Balinsasayao NP
43 Mainit Hot Spring Sa
44 Liguasan Marsh GR

Indonesia

45 S. M. Langkat NP
46 Gunung Leuser NP
47 Berbak NP
48 Sumatera Selatan NP
49 Udjung Kulon-Panaitan NP
50 Way Kambas NP
51 Kotawaringin/Sampit NP
52 Padang Luwai Res
53 Tangkok-Batuangus NC
54 Penandjung NP
55 Nusa Barung NC
56 Ardjuna – Lalidjiwa Res
57 S. M. Bali NP

Israel

131 Hula Lake NR
132 En-gedi NR
133 Elath Gulf NR

Seychelles

221 Mahé, Cousin, Frigate Is

Malaysia

58 King George V NP

Cambodia

59 Angkor Wat NP

Thailand

60 Erewan Waterfall NP
61 Khao Yai NP

Burma

62 Pidaung GR

East Pakistan

63 Sundarbans GSa
64 Chittagong Hills Tract

India

65 Kaziranga GR
66 Manas GR
67 Jaldapara GR
68 Corbett NP
69 Keoladeo-Ghana Sa
70 Kanha NP
71 Gir Forest Res
72 Mudumalai GR
73 Vedanthangal Sa
74 Ranganthittoo Sa
75 Periyar Sa

Ceylon

76 Wilpattu NP
77 Gal Oya NP
78 Ruhuna NP

Madagascar

213 Lokobé Sa
214 Cape Masoala Sa
215 Ankarafantsika Sa
216 Betampona Sa
217 Antsingy Sa
218 Andringitra Sa
219 Tsimanampetsotsa Sa
220 Andohalela Sa

North and western Africa

145 Bou Hedma Sa (Tunisia)
146 Tazekka NP (Morocco)
147 Toubkal NP (Morocco)
148 Niokolo Koba NP (Senegal)
149 Boucle de Baoulé NP (Mali)
150 Nimba Mts NC (Guinea)
151 Mole GR (Ghana)
152 W du-Niger NP (Upper Volta, Niger, and Dahomey)
153 Waza NP (Cameroun)
154 Benoué GR (Cameroun)
155 Faro GR (Cameroun)
156 Boubandjidah GR (Cameroun)
157 Manda GR (Chad)
158 Bamingui-Bangoran NP (CAR)
159 Nana-Barya GR (CAR)
160 Zemango GR (CAR)
161 Odzala NP (Congo, Brazz.)
162 Mont Fouari NP (Congo, Brazz.)
163 Albert NP (Congo, Kinsh.)

Sudan

164 Dinder NP
165 Nimule NP

Ethiopia

166 Simien Mts NP
167 Menagasha NP
168 Awash NP

Uganda

169 Kidepo NP
170 Murchison Falls NP
171 Kigezi Gorilla Sa
172 Kigezi Gorilla Sa

Kenya

173 Aberdare NR
174 Tsavo NP
175 Amboseli NP
176 Nairobi NP
177 Mt Kenya NP
178 Masai Mara GR

Tanzania

179 Serengeti NP
180 Arusha NP
181 Ngurdoto Crater NP
182 Selous CA
183 Gombe Stream GR

Malawi

184 Nyika NP
185 Kasungu NP
186 Lengwe NP

Mozambique

187 Gorongoza NP

Zambia

188 Sumbu GR
189 Luangwa Valley NP
190 Kasanka GR
191 Kafue NP

Rhodesia

192 Mana Pools GR
193 Chewore GR
194 Inyanga NP
195 McIlwaine NP
196 Ngezi NP
197 Matopos NP
198 Wankie NP
199 Victoria Falls NP

South Africa

200 Kruger NP
201 Mkuze GR
202 Hluhluwe GR
203 Umfolozi GR
204 St Lucia BSa
205 Natal NP
206 Loteni NR
207 Willem Pretorius GR
208 Mountain Zebra NP
209 Addo Elephant P
210 Bontebok NP
211 Kalahari NP

South West Africa

212 Etosha Pan Sa

Index

References in *italics* are to illustrations or captions to illustrations. References in **bold** are to map keys.

A

aardvark, **72**, **77**, 82, *82–3*
aardwolf, 72
acouchi, 59
addax (antelope), **32**, 39, **72**
adder, puff, **33**, 39, **73**, **77**
aestivation, 39
Africa, 14, 68–83
agouti, 59
albatross, **106**, 107
Alexander II of Russia, 29
alligator, American, **48**, 51, *51*
alpaca, 66
Altay Mountains, 31
Amazon River, 54
amphibians: in Gondwana, *18*, 20; in Palaearctic, 28
anaconda, 62
Andes Mountains, 19, 65
anoa, 93
Antarctica, 20, 104–7
anteaters, 20, 52, 55; dwarf, 55; giant, **57**, **61**, 63, *63*, **64**; marsupial, *see* marsupial anteater; spiny, *see* echidna; *see also* tamandua
antelopes, 78, 80, 81, 90; four-horned, 86; roan, 82; royal or pygmy, 78; sable, 82; saiga, 22, **33**, 34, *34*, **36**; Tibetan (chiru), 84; *see also* addax, blackbuck, bongo, bushbuck, eland, klipspringer, lechwe, nilgai, nyala, oryx, pronghorn, sitatunga, springbok, steinbok
apes, 75, 91; Barbary, **32**, 35; black, 93; *see also* chimpanzee, gibbon, gorilla, orang-utan
Appalachian Mountains, 45
Arabia, 38
Archaeopteryx, fossil bird, 62
Arctic Ocean, 44
argali (wild sheep), 84
armadillos, 20, 40, 52, 63; fairy, 63; giant, 55, *55*, **57**, **61**, 63, **64**; nine-banded, **49**, **57**, **61**, **64**; three-banded, 63
asses, wild, 24, 31, **36**, 39; African, 70, **73**, **77**; in Australia, 102; Indian, 86, **89**
Atlas Mountains, 30
aurochs, 24
Australian region, *12*, *14*, **17**, 20, 94–103
axolotl, 51
aye-aye, **112**, 113–14, *113*

B

babirusa, 93
baboons, 71, **73**, **77**; gelada, 74, hamadryas, 74

badger, American, 46
bandicoot, **99**
banteng, 90
barbet, 90
bats, 14, 30 91; in Australia, 102; in Galápagos Islands, 116; in New Zealand, 108; short-tailed, **110**; vampire, 55, **57**, 58, **61**, **64**
bears: black, **42**, 44–5, **48**; brown, **27**, 28, 30, **32**, *40–1*, **42**, 44, **48**; Himalayan black, **36**, 84, **89**; polar, 14, 24, 25, **27**, **36**, **43**, 44; sloth, 87, **89**; spectacled, **57**, **61**, 65; sun (or honey), **89**, 91
beavers, **27**, 28, **32**, **36**, **42**, *44*, 45, **48**; mountain (sewellel), 47
bee-eater, 38
Bering Strait, land bridge at, 14, 20, *21*
Bialowieza Forest, bison in, 29
birds: on Krakatau, 117; migratory, 25, 35; in New Zealand, 108, 110; in South America, 52
birds of paradise, 93, **99**, 103; Count Raggi's, *102–3*
bisons, 14; American (buffalo), *21*, 42, 46, *47*, **48**; European (wisent), 24, **27**, 29, **33**; "Indian," 86
blackbuck, 86
blue-buck, 68
boar, wild, 24, **27**, 29, 30, **33**, **36**, 38, 86
bobcat, **42**, 47, **48**, 50
bongo (antelope), 78
booby, 67
boomslang, *80*
bower bird, 103
brockets (deer), red and brown, 59
budgerigar, 103
buffaloes: African, 71, **72**, **77**, 80; Asiatic, 86, 87; forest, 78; water, 90, 102; *see also* bison, American
bush-babies, 78, 90; Senegal, 82
bushbuck, 78
bushveldt, 72
bustard, great, **27**, **33**, 34, *34*, **36**

C

cacomistle, 50
caiman, 62
Callithricidae (monkeys), 58–9
Camargue nature reserve, 35
camels, 39; in America, 40; in Australia, 102; Bactrian, 31, **36**
campos (plains), 62
Canada, 42–3
Canute, King, 29
capybara, **56**, *58*, 59, **61**, 63, **64**
caracal, 70–1
caribou, *22*, **42**, 44
cassowaries, **99**, 103; Australian or common, 100–1; Bennett's, *100*; one-wattled, *100*

cats, wild, **27**, 30, **33**, 34, **36**; African, 86; Australian native, *see* dasyure; golden, 91; jungle, 86, 91; leopard, 90; marbled, 91; mountain, 66; pampas, 65
Catherine the Great of Russia, 28
Caucasian Reserve, *31*
Caucasus Mountains, 30
cavy family, 63
Cebidae (monkeys), 58
Central America 56–7
Chaco, 62–3
chameleons, 35, 79, 114; African flap-necked, *78*
chamois, **27**, **32**
cheetah, 70, **72**, **77**, *80–1*, 81
chevrotains, *12–13*, *14*, 87, **88**; water, 78
chimpanzee, **72**, 75, **77**
chinchilla, **61**, 65
chinkara (gazelle), 86
chipmunk, 45, 46
chiru (Tibetan antelope), 84
chital (deer), 86, 87
civets: African, 78; Oriental, 86, 91; otter, 92–3
climate, and distribution of animals, 14
coati, **48**, 50, **57**, **61**, 62, **64**
cobra, 90, *90*
cock of the rock, 62
cockatoo, sulphur-crested, *13*, *14*
colugo (flying lemur), **88**, 91, *92*, *93*
condor, giant, **57**, **61**, **64**, 66
Continental Drift, theory of, 15, 19–20
convergence, 97
cormorant, 67, 79; flightless, **116**
Coto de Doñana nature reserve, 35, 38
coyote, **43**, 45, 46, **48**, 50, 65
coypu (nutria), 51
crane, whooping, **43**, *44–5*, 45, **49**
crocodiles: African, **72**, **77**, 79; American, 51, 62, 115; Australian, **99**, 102; Cuban, **114**, 115; fish-eating, *see* gavial
crossbill, 28
curassow, 62
cuscus (possum), 93, 97, **99**

D

Darwin, Charles R., 12, 116
dasyure (Australian native cat), *13*, *14*, **99**, 102
deer, 24; barking (muntjac), 87; chital, 86, 87; fallow, **27**, 29, **33**, 35, **36**; marsh, 63; mule, **43**, 46, **49**; pampas, **61**, 63, **64**; red, **27**, *28–9*, 29, **33**, **36**, 38; roe, **27**, 29, **33**, **36**; sambar, 86, 87; sika, **36**; wapiti, *see* elk; white-tailed, **43**, 44, 45, **49**, 50, 59; *see also* brockets, pudu, reindeer

121

deserts, 38–9, 50, 66, 83
dingo, 101
dogs, wild: African, 81, 87; Asiatic, 87,
   90; bush, 62; dingo, 101
dolphin, 107
doves, 71, 90
drill, 75
ducks, 38, 45
dugong, 51, **72**, **77**, **88**, **99**, 102
duikers, 78; red, **73**, **77**

E

eagles, 30, 34, 38; monkey-eating, 93;
   wedge-tailed, 100
echidna (spiny anteater), 96–7
eel, electric, 52
egrets, 38, 51
elands, 78, 82; giant, 71, **73**, **77**
elephants, 21; African, **72**, **77**, 81, 82,
   86–7; forest and bush, 78; Indian or
   Asiatic, 86, *86–7*, 87, **88**, 91
elk, 24, 25, **27**, 28, **36**, **43**, 47, **49**
emu, **99**, 103
ermine, 25, 45
Ethiopian region, **17**, 20, 68–83
Europe: North, 26–7; South, 32–3
evolution by natural selection, theory of,
   12, 14

F

falcon, prairie, 46
finches, Darwin's, 116
fish, in Gondwana, *18*, 20
fishers (weasel family), 44, 45
flamingos, 38, 51; greater, **33**, *34–5*;
   lesser, *35*
flight, spread of animals by, 14
forests, 22, 24, 44; coniferous, 25, 28,
   44; deciduous, 28, 45; tropical, 52, 54,
   75, 84, 87, 91
fowls, *see* guinea fowl, jungle fowl,
   mallee fowl, pea fowl
foxes, 86, 90; Arctic, *24*, 25, **27**, **36**, **42**;
   bat-eared, 71; fennec, **33**, *38–9*, 39;
   introduced into Australia, 102; kit, **49**,
   50; pampas, 65; in suburbs, 30, 45
frigate bird, 67
frogs, 28; flying, 93; New Zealand
   (Leiopelma), 15, 110, **110**; tree, 79
fulmar, 67, 107
fur trade, 28, 30, 44, 45, 59–60, 66, 75,
   100

G

Galápagos Islands, 108, 116–17
galinule, 51
gaur ("Indian bison"), 86, **88**, 90
gavial (gharial), 86, **88**, *92*
gazelles: Arabian, 70; dama, 70;
   giraffe-necked, 71; Grant's, **73**, **77**,
81; Indian, 86, 90; Thomson's, *80–1*,
91
gecko, 35, 83
gemsbok (common oryx), 83
genet, 35, 78
gerbil, 31, 39
gerenuk (giraffe-necked gazelle), 71
gibbon, 75, **89**, 91–2
gila monster, **49**, 50, *50*
giraffe, 20, *70*, 71, **73**, **77**, 81, 82
gnu (wildebeest), 80, 82
goats: domestic, 35; Rocky Mountain,
   43, 47, **49**; wild, 84
Gobi Desert, 38, 39
Gondwana, *18*, 19, 20
gopher, plains pocket, **43**, 46, **49**
goral, 31
gorilla, 75, **77**; mountain, 75; lowland, *74*
guanaco, **61**, **64**, 66
guenon (monkey), 75
guinea fowl, 79
guinea pig, 59, 63, 65
gulls, 44, 45, 67

H

hamster, common and golden, 31, 34
hares, 25; whistling (pika), 47
hartebeest, 71, 80, 82
hawks, 50
hedgerows, 29
hemione (wild ass), 86
herons, 51, 79
hibernation, 31, 34
Himalayas, 19, 84
hippopotamuses, 20, **73**, **77**, *78–9*, 80,
   82; pygmy, 78
hoatzin, 62
Holarctic region, 16, 40
honeyguide, 83
hoopoe, 38
hornbill, **73**, **77**, 79, 82; yellow-billed, *80*
horses, wild: in America, 40; in Australia,
   102; grey tarpan, 22; Przewalski's, 22,
   31, *31*, **36**
hummingbirds, 51, **57**, **61**, 62, **64**, 66;
   sicklebill, *62*
hutia, **114**, 115
hyena, **33**, 39, *69*, **73**, **77**, 81, 90
hyrax, 78; rock, 75, 83

I

ibexes, **27**, 30, **33**, 34, **36**; Abyssinian,
   72, **77**; Alpine, *30*; walia, 74
ibis, 79
Ice Age, 14, 20–1; Pleistocene, *21*
iguanas: land, 116, *116*, **116**; marine,
   116, **116**
impala, **73**, **77**, 80, 82
India, 86–90
Indo-China, 90–2
indri (lemur), **112**, 113, *113*
insects: in Africa, 68; in tropical forest, 54
Island regions, **17**, 108–17

J

jackal, **27**, 30, **33**, 39, **73**, **77**, 81, 86, 90;
   Simenian (Abyssinian wolf), 74
jaguar, **49**, **57**, 59, **61**, *62*, **64**, 65
jaguarundi, 50, 59
jerboa, 31, **33**, *38–9*, 39; marsupial, *see*
   marsupial jerboa
jungle fowl, **89**, 90, **99**

K

kakapo (owl-parrot), 110, **110**
Kalahari Desert, 83
kangaroos, 20, 100–1; grey, 94–5, **99**;
   rat, **99**, 100; red, *94–5*, 95–6, **99**, 101;
   tree, **99**, 101; white-throated tree, *13*,
   14
kea (parrot), 110, **110**
kinkajou, **57**, **61**, 62
kite, 34
kiwi, 108, 110, **110**
klipspringer (antelope), 75
koala, 14, 20, **99**, 100
Komodo dragon, **99**, 103
kouprey, 90
Krakatau, recolonization of, 117
krill, 107
Kruger National Park, 82
kudus (antelopes): greater, 82, *82–3*;
   lesser, 81
kusimanse (mongoose), 79

L

langur (monkey), 86, **89**
larks, 34, 71
Laurasia, *18*, 19, 20
lechwe (antelope), **77**, 79
lemmings, *24*, 25, 44; brown and
   collared, **27**, **36**, **43**
lemurs, 113; mouse, **112**; ring-tailed,
   112, 113
leopards, **33**, **36**, **73**, **77**, 78, 81, 86, 87,
   89, 90, 91; Barbary, 30; clouded, 91;
   snow, 31, **36**
lions, 34, *68–9*, **73**, **77**, 81; Asiatic, 86,
   88
lizards: in Australia, 102; beaded, 50;
   in deserts, 66, 83; flying, **88**, 93, **99**;
   frilled, 102; in Madagascar, 15, 114;
   in Mediterranean lands, 35; spiny-
   tailed, 39; on steppes, 34; *see also*
   gila monster, iguanas, monitors
llama, 66
llanos (plains), 62
lorises, 91; slender, **89**, 90; slow, 90
lungfish, 14, *18*, 52, 54, 102
lynxes, 24, **27**, 28, 30, **33**, 34, **36**, **43**, 44,
   47, **49**; caracal, 70–1; Mediterranean,
   or pardel, *35*, 38; Spanish, *35*
lyrebird, **99**, 103

m

macaques, **36**, 86, **89**
macaw, 62
Madagascar, 15, 108, 112–14
Malay Peninsula, 92
mallee fowl, **99**, 103
mamba, 79
mammals: edentate, 52, 55; evolution of, *18*, 20; monotreme, 96; placental and marsupial, 94
mammoth, 21
man, and distribution of animals, 21, 22
manatee, 51; Amazonian, **61**
mandrill, 75
mangabeys, 75
mara (Patagonian cavy), 63
marabou (stork), 81
markhor (wild goat), 84
marmosets, **57**, 58–9, **61**, **64**
marmot, **27**, 30, **33**, 34, 47
marsupial anteater (numbat), 97, **99**
marsupial jerboa, 97, **99**
marsupial mole, 97, **99**
marsupial mouse, *96*, 97; fat-tailed **99**
marsupial rat, 97
marsupial wolf (thylacine), *100*, 102
marsupials, 94
marten, *28*, 44
matamata (turtle), 62
medicine, Chinese: animal products in, 31, 34
Mediterranean lands, 34, 38
meerkat, 83
mice, 102; field, 29; Galápagos, 116; harvest, 30; house, 30; introduced into West Indies, 115; pocket, 50; scaly-tailed, 78
Mid-Atlantic Ridge, 15
moa, 110
mole: marsupial, *see* marsupial mole; star-nosed, 45
mole-rat, 34, 81
mongoose, 35, 78; Indian grey, 86, **89**, 90, *90*; introduced into West Indies, 115
monitors (lizards): Australian, 102, 103; Nile, 79
monkeys, 21, 58–9; capuchin, 56, 58, **61**, **64**; Celebes, 93; colobus, **73**, 75, *75*, **77**; douroucouli (owl, or night), 58; forest-living, **73**, **77**; grassland, **73**, **77**; howler, **56**, 58, **61**, **64**; patas, 71; proboscis, 93; Roxellane's or golden, 86; saki, 58; spider, **56**, 58–9, *59*, **61**, **64**; squirrel, 58; titi, 58; uakari, 58, 59; vervet, 71; *see also* baboons, drill, guenon, langur, macaques, mandrill, mangabeys, marmosets, tamarins
monotremes, 96
moon rat, Malayan, **89**, 92–3
moose, *21*, **43**, 44, **49**
mouflon (wild sheep), **33**, 35
mountain devil (lizard), 102
mountains, as refuge for animals, 22, 30, 47
mouse, *see* mice
muntjac (deer), 87
musk ox, *21*, 24–5, **27**, **36**, **43**, 44
muskrat, 51
myna, 90
myxomatosis of rabbits, 102

n

Nearctic region, 14, **17**, 20, 21, *21*, 40–51
Neotropical region, 15, **17**, 20, 52–67
New Zealand, 100, 108–11
newts, 28, 45
nilgai (antelope), 86, **89**
numbat (marsupial anteater), 97, **99**
nutria (coypu), 51
nyala (antelope), 82

o

ocelot, 50
okapi, *70*, **73**, **77**, 78
olm, 28, *28–9*, **33**
opossums, 40, **43**, 45, **49**, 52, 54, **57**, **61**; mouse, *54*; Virginian, 54–5; woolly, 54
orang-utan, 75, *84–5*, **89**, 93
oriole, 62
Oriental region, *12*, *14*, **17**, 84–93
oryxes, **73**; Arabian, 39, 70; common (gemsbok), 83; scimitar-horned, 70
osprey, 51
ostrich, *70–1*, 71, **73**, **77**, 81
otter civet, 92–3
otters, 45, 51; giant, **57**, 59, **61**; sea, **43**, 44, **49**
oven birds, 63, 65, 66
owl-parrot (kakapo), 110
owls, 28, 45; burrowing, 63; elf (or cactus), 50; snowy, *24–5*, 25, 44

p

paca, 59, 65
Palaearctic region, 14, **17**, 20, 21, *21*, 22–39
pampas (plains), 62
pandas, 14, *30–1*, 31, 84; giant, 84, **89**; red or lesser, 84, 86
Pangaea, *18*, 20
pangolins, **73**, **77**, 78; African tree, 78; Indian, **89**, *90–1*, 91
panther, 91
parakeets, 90
parrots, 51, 79, 103; *see also* kea
pea fowl: Congo, 79; Indian, **89**, 90
peccary, **57**, 59, **61**, *62*, **64**
pelicans, 79; brown, 67
penguins, 67, 104; Adélie, 104, **106**; emperor, 104, *104–5*, **106**, 107; Galápagos, 67, **116**, Magellanic, 67; Peruvian, **61**, **64**, 67
pesticides, 24
pet food, kangaroo meat for, 101
pet trade, 59
petrels, 67, 107
phalangers (possums), 97, 100; flying (gliders), 97, **99**, 100; honey, 97; lesser gliding, *97*; *see also* cuscus
pheasant, Lady Amherst's, *13*, *14*
pigeons, 103; crowned, **99**, *101*

pigs, wild, 14; in Australia, 102; forest, 78; *see also* babirusa, boar (wild), warthog
pipits, 34, 71
piranha fish, 52
plankton, 66
platypus, duck-billed, 96, *96*, **99**
polar bear, *see under* bears
polecat, 34
pollution of water, 24
porcupines, **43**, 45, **49**; crested, **32**, 35
porpoise, 107
possums, *see* phalangers
potto, **73**, **77**, 78, 90
prairie chicken, **43**, 46, **49**
prairie dog, **43**, *46*, **49**
prairies, 46
pronghorn (antelope), 40, **43**, 46, *46–7*, **49**
ptarmigan, 25, 44
pudu (deer), **61**, **64**, 65
puma, **43**, 47, **49**, **57**, 59, **61**, 65, 66

r

rabbits: in Australia, 102; long-eared jack, 50
raccoons, **43**, 45, *45*, **49**, 51; crab-eating, 62
ratel (honey badger), 83
rats, 30, 78, 102; introduced into West Indies, 115; kangaroo, **49**, 50; pack, 50; marsupial, *see* marsupial rat; whistling, 83; *see also* moon rat
rattlesnake, sidewinder, **49**, 50, *50*
reindeer, *22–3*, 25, **36**
reptiles, 14; in deserts, 39, 50; in Gondwana, *18*, 20
rhea, **64**, 65, *65*
rhinoceroses: black, **72**, **77**, 80, 81, *81*, 82; Javan, 93; one-horned Indian, 86, **89**; Sumatran, 91; white, **73**, **77**, 80, 82
Rocky Mountains, 47

s

sable, 28
Sahara Desert, 38, 39, 68
salamanders, 28, 45; axolotl, 51; hellbender, 45; Japanese giant, 28, **36**
sambar (deer), 86, 87
savannas, 70, 82, 90
Sclater, P.L., 12
scorpion, 50
sea lions: Australian, 102; Southern, **61**, **64**, 67
seals, 24, 44; Australian fur, 102; Baikal, **36**; common, **27**, **32**, **36**, **43**, 48; crabeater, **106**, 107; elephant, **106**, 107; grey, **27**, **36**, **43**; leopard, 104, **106**; Mediterranean monk, **33**, 35, **73**; northern fur, 44; Ross, **106**, 107; southern fur, **61**, **64**, 67, **77**; Weddell, **106**, 107, *107*; West Indian monk, 114, *115*
secretary bird, **73**, **77**
Serengeti National Park, 80
serow, 31

123

sewellel (mountain beaver), 47
sheep, wild; Barbary (aoudad), 30, **32**;
   bighorn, **42**, 47, **48**; *see also* argali,
   mouflon
shoebill, **77**, 79, *79*
shrews, 78; long-nosed elephant, 78
siamang (ape), 92
sifaka (lemur), **112**, 113
sitatunga (antelope), 79
skua, 107
skunk, 46, 50
sloths, 20, 52, 55, **57**, **61**, **64**; three-
   toed, *52*; two-toed, *52–3*
snakes, 45, 90; constricting, 52; coral,
   50, 62; paradise tree, 93; on steppes,
   34; tree, 51; water, 51
solenodon, 114, *114–15*, **114**
South America, 14, 52–67
sparrow, house, 30
spiders: black widow, 50; tarantula, 50
spoonbill, 28, 79
springbok (antelope), 83
springhaas, **77**
squirrels, 45, 87, 91; flying, 87, 91;
   ground, *46*, 83; Malabar, 87; 90;
   scaly-tailed flying, 78; *see also* suslik
starling, 31, 34
steinbok (antelope), 75
steppes, 31
stoat, introduced into New Zealand, 110
storks, 79; marabou, 81
suslik (ground squirrel), **27**, 31, **33**, 34,
   36
swallow, 30
swans: trumpeter, 45; whistling, 45;
   white mute, 45
swimming, spread of animals by, 14

tahr (wild goat), 84
taiga, 25, 28
takahé, *108–9*, 110, **110**
takin, 84
tamandua, 55, **57**, **61**, **64**
tamarau, 93
tamarins, **57**, 58–9, **61**, **64**
tapirs, 14, 21, 52, 59; Brazilian, **56**, **61**,
   64; Malayan, or Asiatic, **89**, 92, *92*
tarpan (horse), 22
tarsier, *12–13*, *14*, **89**, 91, 93
Tasmanian devil, **99**, 102
tenrecs, **112**, 113
termites, 55, 72
terns, 67; Arctic, 44

Tethys Ocean, 19, 20
thylacine (marsupial wolf), *100*, 102
tigers, 30–1, **36**, 87, **89**, *90–1*, 90, 91;
   Javan, 93; sabre-toothed, 21
tinamou, 65
toads, 28; ant-eating, 52; horned, 50
tortoises, 14; giant, 116–17, *116–17*,
   116
toucan, **57**, **61**, 62, **64**
touraco, 79
transitional zones, **17**, 93
tree-shrews, *12*, *14*, **88**, 91, *92*, *93*
tsetse fly, 68, 82
tuatara, 110, *110*, **110**
tuco-tuco, 63
tundra, 25, 44
turkeys, wild, 40, 45–6, **49**; brush, 103
turtles, 45, 51, 62; alligator snapper, **48**,
   50–1, 51; side-necked, 14, *18*, 54, 102

Udjung Kulon Reserve, Java, *87*
umbrella bird, 62
United States of America, 48–9

vegetation, and distribution of animals. 14
vicuña, **61**, **64**, 66, *66–7*
vipers: gaboon, **73**, **77**, 79; green tree,
   79; horned, 39; pit, 62, 90; side-
   winding, 83
viscacha, 63, **64**, 65
volcanoes, 19
vole, 25, 29
vultures, 30, 38, 81

wallaby, **99**, 100
Wallace, A. R., 12, *12*, 108
wallaroo, **99**
walrus, 24, 44
warthog, 71, **73**, **77**
waterbuck, 80
weasel, 44; introduced into Australia, 102
weaver birds, 71, 82
Wegener, A., 15
West Indies, 108, 114
whales, 44; blue, fin, humpback,
   sperm, 107; killer, *104–5*, **106**, 107;
   toothed and whalebone, 107
wildebeest (gnu), 80, 81, 83
wolverine (glutton), 25, **27**, 28, **36**, **43**,
   44, 45
wolves, 24, 25, **27**, 28, 30, **33**, **36**, 38,
   43, 44, 86, 90; Abyssinian
   (Simenian jackal), 74; grey or
   timber, *25*, 45; maned, **61**, **64**, 65, *66*
wombat, **99**, 100; coarse-haired, *97*
woodpeckers, 28, 50
wrens: cactus, 50; flightless, 110

yak, 84, **89**
Yellowstone National Park, 40, 46, *47*

zebra, *68–9*, **73**, **77**, 80, 81, 82, 83;
   Grevy's, 71
zoogeographical regions, 12, 16–17

Acknowledgments:

Page 18 after Robert S. Dietz and John C. Holden, *The Breakup of Pangaea,* © October, 1970 by Scientific American, Inc. All rights reserved. Page 118–119 courtesy *The Advanced Atlas,* John Bartholomew & Son Ltd., Edinburgh, 1962.